Open Divide

Open Divide
Critical Studies on Open Access

Edited by

Ulrich Herb
University and State Library,
Saarland University (Germany)

Joachim Schöpfel
Department of Information and Document Sciences,
University of Lille (France)

Library Juice Press
Sacramento, CA

Published by Library Juice Press in 2018

Library Juice Press
PO Box 188784
Sacramento, CA 95818

http://libraryjuicepress.com

Library of Congress Cataloging-in-Publication Data

Names: Herb, Ulrich, editor. | Schèopfel, Joachim, 1957- editor.
Title: Open divide : critical studies on open access / edited by Ulrich Herb, University and State Library, Saarland University (Germany); Joachim Schèopfel, Department of Information and Document Sciences, University of Lille (France).
Description: Sacramento, CA : Litwin Books, 2018. | Includes bibliographical references and index.
Identifiers: LCCN 2018001495 | ISBN 9781634000291 (acid-free paper)
Subjects: LCSH: Open access publishing. | Open access publishing--Developing countries. | Communication in science. | Communication in learning and scholarship.
Classification: LCC Z286.O63 O66 2018 | DDC 070.5/7973--dc23
LC record available at https://lccn.loc.gov/2018001495

Contents

Preface

Richard Poynder

When the internet emerged open access to publicly-funded research appeared to be a no-brainer. The network, it was argued, could dispense with scholarly journals' print and postage costs and allow papers to be shared more quickly, more cost-effectively, and in a way that would level the playing field for those in the developing world—since it would be possible to make articles freely available on a global basis. As a result, the 2002 Budapest Open Access Initiative (BOAI) declared, the research community would be able to "share the learning of the rich with the poor and the poor with the rich ... and lay the foundation for uniting humanity in a common intellectual conversation and quest for knowledge."

As proof of concept, open access advocates pointed to arXiv, the online preprint server that physicists have been using to share their papers since 1991.

But while the potential benefits of open access are undeniable, making it a reality has turned out to be a slow and difficult process, and it remains far from clear that it will lead to an inexpensive or levelling way of sharing research.

It turns out, for instance, that researchers are a surprisingly conservative bunch, a characteristic reinforced by the promotion and tenure (P&T) systems that operate in academia. Consequently, most authors have continued to share their work in the traditional manner using traditional publishers, and in ways that reinforce the traditional hierarchical and elitist culture that has prevailed in the research community since time immemorial.

Publishers were also initially cautious about open access—amply demonstrated in 1999, when the then director of the US National Institutes of Health, Harold Varmus, proposed the creation of E-Biomed. Intended to replicate and extend the arXiv model in the biomedical field, Varmus' plan envisaged a biomedical preprint server and new electronic journals managed by an E-Biomed governing body. It also assumed that authors would retain copyright in their works, a proposal that, in itself, was enough to give publishers the jitters.

Unsurprisingly, therefore, publishers responded to the E-Biomed proposal with doomsday predictions about the imminent collapse of the scholarly communication system and intense political lobbying. This saw Varmus' proposal significantly watered down and launched as PubMed Central in 2000. Gone was the preprint server, gone were the new journals, and gone was the expectation that authors would retain copyright. Gone also was what had, in essence, been an attempt by the research community to wrest control of scholarly communication from legacy publishers. For the open access movement this was a significant defeat.

But advocates persisted in their calls for open access, and publishers had eventually to conclude that they could not hold the tide back indefinitely. Fortuitously for them, new-style open-access publishers like Public Library of Science (co-founded by Varmus) and BioMed Central (subsequently acquired by legacy publisher Springer Nature) had by then demonstrated that it is possible to fund open access by levying publication fees in place of subscriptions (i.e. offer pay-to-play gold OA). Incumbent publishers realised that if those fees were set high enough they could embrace open access without any diminution of their substantial profits. So, they began to launch their own open access journals, and to introduce hybrid open access, which allows researchers to continue publishing in traditional journals *and* make their papers open access—so long as they pay a premium (c. $3,000 per paper).

However, some open access advocates pointed out that gold open access would unnecessarily enrich publishers at the expense of the research community, not least because hybrid open access provides publishers with an additional, rather than a replacement, revenue stream—i.e. subscriptions *and* publishing fees. As such, they suggested, researchers should continue publishing in subscription journals without paying a fee, and then self-archive copies of their papers in their institutional repositories, and in this way make them freely available to all—aka green open access. Attracted by this more cost-effective approach, funders and institutions began to introduce open-access policies requiring researchers to self-archive—with, it has to be said, limited success, since most researchers simply ignored the policies.

Seeing green open access as a direct threat to their revenues, publishers began imposing ever more lengthy embargoes and ever more complex and onerous rules over when, where, and what version of a paper can be made open access. They could do this because—as a condition of publication—authors are required to assign copyright in their work to the publisher. The consequent complexity of green open access served to strengthen researchers' resistance to self-archiving, and today green open access looks like a failed strategy. Gold open access, by contrast, has gained considerable traction.

A key moment came in 2012, with the publication of the Finch Report. Produced by a UK government-appointed committee overrepresented by publishers, Finch concluded that pay-to-play gold open access was the best approach, not least because it protected publishers' existing revenues. It was at this point that publishers began to co-opt open access—a development amply aided by the fact that open access advocates were by now thoroughly divided over how to achieve open access, or even exactly what it is. As a result, funders and governments began to turn to publishers for direction more often than to the representatives of the open access movement.

Thus, in the wake of Finch, other national and international initiatives have emerged that also prioritise gold open access. In 2016, for instance, a number of European funders launched the OA2020 initiative "to convert the majority of today's scholarly journals from subscription to Open Access (OA) publishing". The same year the EU called for 'immediate' open access to all scientific papers by 2020 (which inevitably implies gold OA).

The appeal of gold open access is that it is far simpler, and allows the final version of a paper (rather than a preprint) to be made freely available. Moreover, since it means that no embargoes need be imposed papers become immediately available online. Importantly, publishers far prefer gold to green open access. The problem is that gold open access increases rather than reduces the cost of scholarly communication, and so confounds BOAI's expectation that open access will be more cost-effective.

For researchers based in the global South the emergence of pay-to-publish open access is especially troubling. Increasingly incentivised to publish in prestigious international journals (which are invariably based in the global North) researchers in the developing world face the prospect of having to pay publishing fees of hundreds or thousands of dollars every time they need to publish a paper, something few can afford to do.

As such, OA's promise that it would level the playing field has also been confounded. Indeed, open access now looks set to widen rather than narrow the North/South knowledge divide. Consider, for instance, that BOAI assumed that if a paper was made open access it would be free for anyone to access. Elsevier's response to European calls for subscription journals to be converted to gold open access, however, has been to propose what it calls "region-specific OA". This envisages that access to papers would be granted or denied depending on a researcher's geographical location, with access limited to residents of the country/region that has paid the cost of publication. This, of course, cannot fairly be described as open access. Rather it is (counter-intuitively) an open access version of the toll access national licensing schemes that organisations like the UK's Jisc regularly negotiate.

Elsevier's idea may come to nothing, but that such a thing as regional open access could be proposed draws our attention to the fact that—far from being inclusive—open access may further disenfranchise those in the global South. After all, Elsevier estimates that 80% of papers are still published toll access. This means that researchers in the global South now face a double barrier. To provide faculty with access to the 80% of research behind paywalls institutions in the developing world would need to pay subscription fees, but few can afford to subscribe to more than a handful of journals. This is the historic toll access barrier.

In addition, as journals start to flip to gold open access researchers in the developing world will discover that they cannot afford to publish their own research. This is a new barrier, and a direct consequence of the demands for open access; a barrier, moreover, that will exclude researchers in the global South from the "common intellectual conversation" promised by BOAI.

To cap it all, gold open access has unleashed on the world a plague of predatory journals, with those in the South said to be disproportionately impacted.

Meanwhile, anyone whose first language is not English faces a language barrier too, since English has become the *lingua franca* of scholarly communication. In addition, those without adequate internet access face a bandwidth barrier. These are not barriers that were addressed at BOAI, but they need to be taken into account when discussing open access.

Of course, researchers in the global South have the option of spurning international journals and making their work freely available in their own language, in a local repository. But this cannot provide the visibility that publishing in an international journal can, and it will not satisfy their employers' P&T requirement that they publish in prestigious journals.

In short, while open access promised to create a cheaper, faster, and more inclusive system of scholarly communication, it now seems likely to be more expensive and to widen the North/South knowledge divide. Indeed, some believe that open access could prove a new source of colonialism, with scientists in the North able to freely plunder knowledge produced in the South while continuing to define and control what counts as scientific knowledge, and who can contribute to it. Those in the developing world will still be locked out of the conversation.

Clearly, if the BOAI promises are to be met the current trajectory of open access would need some adjustment. Two developments might appear to hold out some hope.

First, there is growing interest in so-called diamond open access, in which journals charge neither publication fees nor access charges. Costs are covered by other means—through sponsorship by a learned society, for instance, through endowments, or by means of government grants. Open access advocates frequently cite as a model here SciELO—the publicly-funded Latin American co-operative publishing platform. SciELO, they point out, offers a cheaper alternative to the model emerging in the North (SciELO costs are estimated at $90 per article). Given the traction that pay-to-publish has now acquired, however, diamond open access could struggle to gain mindshare.

The second development to note is the reinvigorated preprint movement. As I write this, new services like bioRxiv, SocArXiv, EarthArXiv, and PsyArXiv are emerging on an almost weekly basis. It is worth noting that bioRxiv is essentially the preprint server Varmus wanted to introduce 18 years ago.

Potentially, preprint servers could deliver on all three open access promises—i.e. provide a faster, cheaper, and fairer system for sharing research. Indeed, in theory, they could make the traditional journal redundant, and so deliver very significant cost savings (it is estimated that it costs just $7 per paper to post and host on arXiv).

On the other hand, papers deposited in preprint servers are invariably later submitted to legacy journals, if only in order to meet the demands of P&T committees. In a gold open access world this would mean authors were still confronted with high publishing fees. So, it is not obvious that preprint servers will deliver the cost reductions that are essential if the developing world is to become an equal partner in the open access world.

We need also to view open access against the backdrop of a larger drive for openness. Not only does open access now encompass monographs as well as research papers (which presents a new set of problems), but we have also seen the emergence of the open data and OER movements, along with the broader open science movement. Looking further out, there are also the commons/commoning movements. All these movements are products of the internet, and they were all initially infused with a belief that some areas of human endeavour should be based on public rather than private goods.

The challenge all these movements face, however, is that we live at a time when neoliberalism—and a belief in the primacy of the market—dominates both public discourse and public policy. What the experience of the open access movement has taught us is that while alternative solutions intended to operate outside the straightjacket of the market are highly desirable (and highly desired), they are difficult to sustain. Public goods are constantly vulnerable to subversion, marginalisation and/or privatisation by commercial interests. It does not help that some open advocates have sought to promote their cause by promising it will provide commercial benefits as much as non-monetary social value. And when it comes to competing in markets the North continues to enjoy inherited advantages.

Further complicating the picture, powerful global companies like Facebook and Google now manage and control much of the information flow on the Web. Amongst other things this means that making content freely available on the internet does not necessarily make it visible. We should not doubt that more and more open access content will become available online, but it will increasingly be swamped by the tide of non-research information flooding the network. Locating relevant material, therefore, will become ever more difficult, and will create a growing need for specialist pay-to-find discovery services. Those wishing to participate in the common intellectual conversation who cannot afford such services will be at a disadvantage.

Finally, we could note that the Web has created the so-called "platform economy", exemplified by for-profit services like Uber and Airbnb. This is the direction that scholarly communication is now taking, with commercial repository services like SSRN, and paper sharing platforms like Mendeley, ResearchGate and Academia.edu leading the charge. Amongst other things, these platforms will aim to capture usage data and sell it back to the research community, with researchers themselves (rather than their research) becoming the product. The implications of this are not entirely clear today, but such services are unlikely to narrow the North/South knowledge divide, not least because the paywalls that the open access movement has spent the last fifteen years trying to pull down look set to be replaced by new ones.

Our conclusion has therefore to be that while most research looks set to become freely available, it is far from clear that open access will level the playing field, or lead to a more cost-effective scholarly communication system. This is unwelcome news for researchers in the global South.

Introduction: Open Divide Emerges as Open Access Unfolds

Ulrich Herb and Joachim Schöpfel

Thirty years have passed since *New Horizons in Adult Education*, the first online peer-reviewed journal, was launched by the Syracuse University Kellogg Project. Fifteen years have passed since the first notable declaration on open access, the Declaration of the Budapest Open Access Initiative (BOAI). It seems it is time to risk a critical review of open access, at a time when open access is being renegotiated.

Open access was at its beginning, if we could date it to the year 1987 or 2001, a rather simple undertaking that knew only one imperative—disseminate scientific publications as quickly as possible without paying any licensing fee. Much has changed since. Whilst *New Horizons in Adult Education* was founded on the initiative of scientists, in 2002 research sponsors and libraries had already discovered open access for themselves and, at the latest with the purchase of the open access publishing house Biomed Central in 2008, also the commercial publishers. But their material interests were opposed: while librarians and funders considered open access as a way to render scientific information less expensive and to make savings on the library budget, publishers learned how to make money with open access, even more money than with the subscription model alone.

Even the definition of what open access could be has become more and more differentiated. While the BOAI was still just trying to distinguish between green open access and gold open access, there are the 2017 terms such as hybrid open access, platinum open access, bronze open access and numerous grey areas of open access, such as when scientists post articles on social networks or publishers provide articles for free online reading without the possibility of printing or downloading them. This confusion corresponds to on-going battles for the definition of supposedly true open access. These discussions are interwoven with licensing issues, which are also reflected in (from the perspective of 1987) unexpectedly subtle distinctions in free, libre or true open access.

However, these differentiations are not mere navel-gazing. They point to an ongoing definitional challenge regarding what open access is and could become.

Indeed, these definitional and policy discussions have substantial real-world consequences. For example, the promotion of paid gold open access by research funders advances the commercialisation of open access and could increase the appeal of dubious publishing practices such as predatory publishing. The current preference for the provision of texts under licences that emulate open source conditions is unsettling for Humanities scholars who are already struggling with the free provision of texts and are also sometimes unfamiliar with open access because there are comparably few attractive open access publication options for books.[1] While open access is generally focused on scientific publishing, which will be the main focus of this book, the issues raised in the discussion of open access touch on all sectors of scholarly publishing.

The upshot is that there is no such thing as one and only understanding of open access, which is more evident today than ever. Open access is differentiated and at the same time integrated into social and economic patterns, interests and developments. These differentiations, as well as the social and economic agendas attached to them, are sometimes accompanied by authoritarian or administrative measures and lead to conflicts. The lines of confrontation are no longer as clear as they were in 2002 when research funders, scientists and libraries formed a coalition against commercial publishers. Today we find that all of these actors form alliances with other stakeholders that are varying and fluid, in arrangements that are both applauded and criticized.

Something has gone wrong along the road to achieving complete open access in scholarly publishing, and the stakes have changed.

---

1 In mid-January 2018, a central directory of open access books and open access book publishers, the Directory of Open Access Books (DOAB, https://www.doabooks.org/), listed 253 publishers, many of whom had published less than ten books.

Thirty years ago, there was a problem with access to information. Today, we have a problem with who controls publishing and moreover, with the control and governance of the whole range of scientific process and output. Former inequalities remain. New inequalities are emerging. Is open access the beginning of a new and more egalitarian episode of scientific communication? Or is it just another Trojan horse, allowing private companies to extend their control of the Big Data now generated by science? Open Divide is dedicated to these conflicts and upheavals in open access. The book has two parts. The first part deals with general questions of open access while the second shifts the focus on open access in the Global South.

At the beginning of the first part, Jutta Haider (Lund) confronts us with the question as to whether open access, driven by economic interests, will become an instrument for external evaluation and control of science. Based on a critical evaluation of recent European initiatives, she states that open access has been turned into an indicator, one amongst many, to motivate researchers and to assess their performance and that a significant part of open access and open science has become a part of a drive towards further economization and privatization of a vast and diverse field of publicly funded knowledge production. New initiatives, disruption and transition are also the topic of the following paper. Elena Šimukovič (Vienna) critically reflects on the dominant discussion of gold open access and argues for plurality of options that might even take—as the author puts it—"deviant directions." Otherwise, she warns, open access might simply replace the "pay-to-read" barrier with a "pay-to-say" barrier.

Samuel Moore (London) shows the different meanings of, expectations of and interests in open access related to the diverse semantics of the concept of openness. There are many diverse approaches espousing the open philosophy, and perhaps this means that 'the open' has a more complicated relationship with the political than meets the eye. He argues that openness does have a real basis outside of neoliberalism and that it is an approach or process that requires careful articulation in an un-decidable terrain. Joachim Schöpfel (Lille) points out that while open access started as a grassroots initiative, it has since undergone fundamental change. Nowadays it is driven by commercial, institutional and political interests and this development may produce dysfunctional patterns, new barriers and unanticipated digital divides. The following paper by Ulrich Herb (Saarbrücken) states that these new barriers (unintentionally and at the same time intentionally) also serve to produce exclusivity and prestige for scientific institutions.

Soenke Zehle (Saarbrücken) gives a constructive and optimistic finale to the first part of Open Divide that deals with critical issues of open access. His text describes economic, social and political strategies for an economy of shared ownership and collective self-organization. He argues for the concept of "cooperative futures" and observes that the interest in various new forms of cooperation are driven largely by similar concerns—namely, searching for ways of collaboration that allow a much higher degree of individual and collective self-determination to pursue shared concerns and to achieve progress towards a "cosmopolitics of the common".

The second part of our book focuses on the implications of open access for the digital divide between the Global North and South. This part opens with an overview on the situation of open access in the Global South by Hélène Prost (Nancy) and Joachim Schöpfel (Lille). Based on figures from directories and databases, they try to answer the question of whether repositories and open access journals change the situation of unequal scientific production. Florence Piron (Montreal) continues this discussion, reporting from Haiti and Francophone Africa where open access can become a tool of neocolonialism if it only provides better access to the science from the North without becoming an instrument of emancipation. She is convinced that African university libraries, if better funded and their staff better trained in digital open access technologies, could play a major role in locating, archiving and preserving local scientific documents as well as in the management of these archives, and she reminds readers that African and Haitian students may have other referents or ideals than the Harvard model. For Piron open access could be a way for researchers to discover the scientific and cognitive heritage of their country in order to gain confidence in their ability to create knowledge relevant to their community. Iryna Kuchma (Vilnius) also sheds light on the impact of open access in the Global South but with particular attention to collaborative open access initiatives. Based on years of practical experience with the Electronic Information for Libraries (EIFL) foundation, her paper is a powerful argument in favor of global networking. Elizabeth Mlambo (Harare) identifies the challenges that the Global South has faced in relation to open access, benefits that have arisen for the Global South and the different approaches taken to open access by the Global North and how these have also benefited from the contibutions of the Global South. She is convinced that open access is now firmly part of the global knowledge creation and dissemination landscape and that African researches can now achieve much needed visibility and reach wider audiences. But she warns that with the Global North still dominating the scene, there is a danger of drowning out scholarship from the Global South if the playing field remains uneven.

Beatriz de los Arcos and Martin Weller from the Open University (Milton Keynes) broaden the perspective of open access to open educational resources (OER) and discuss the extent to which a digital divide between the Global South and North can be identified also in terms of the production and sharing of OER content. Their conviction is that sharing behavior needs to be fostered, facilitated and celebrated by North and South alike, and they call for the discussion to include honest reflection on how much we are taking and how much we are giving back to the cause of open education. Reggie Raju (Cape Town) takes the perspective of the Global South semantically. He differentiates between the social open access components of the Global North, in the form of social justice on the one hand and Ubuntu on the other hand, a Zulu word for advancing communal justice in order to promote an egalitarian society. In his words, Ubuntu, social justice and open access are all part of the same continuum, that is, a continuum towards an egalitarian society—a society that is not compromised by the lack of access to information to meet its development needs.

Our book concludes with an interview with Leslie Chan (Toronto), one of the first signatories of the Budapest Open Initiative Declaration, which gives a resume on open access development and explains Chan's initial interest in open access in the early 90's. This arose not because of the lack of access to scientific information from high-priced subscription journals, but because of the lack of knowledge about high-quality literature from the Global South. The effort to remedy this deficiency led to the launching of Bioline International, a recognised market leader in the provision of open access to peer reviewed bioscience journals published in developing countries. The interview also reports on similar initiatives such as Medknow or SciELO, criteria for their success or failure and the extent to which citation impact and Altmetrics discriminate against scholarly publications from the Global South.

Taken as a whole, all of these contributions aim to present a *panopticon* of open access, which on the one hand sheds light on the inherent and potentially critical nature of open access and on the other hand examines the effects of open access on the exchange of scientific information and privileges between the Global North and South. Obviously this book deals with conflicts in and around open access—conflicts, which also accompanied the work on Open Divide and which led to the fact that not all

11

of the articles initially planned became part of the volume. This applies for example, to an article on "predatory publishing" (Jeffrey Beall) and its impact on open access, as a form of "parody"[2] of the commercial journal publishing itself. Other urgent topics do not receive particular attention: for example, a critical analysis of the economic and financial aspects of open access, or an even deeper analysis of the relationship between open access and evaluation. It seems that already in December 2017 there will be enough material for a continuation of Open Divide into additional volumes.

This profusion of material for new volumes of Open Divide reaffirms the fact that the World Wide Web, with its new technologies and tools and its culture of sharing and common goods, enables academic and research institutions to conduct science in a different way than before. Open science has become the new paradigm of public research. Last year, the member states of the European Union endorsed the so-called *Amsterdam Call for Action on Open Science,* which established two major goals, i.e. that full open access for all publicly-funded scientific publications should be achieved by 2020, and open data should be the standard for all publicly funded research.[3] The European Council published at the end of May 2016 a statement where the Member States agree "that the results of publicly funded research should be made available in an as open as possible manner and (...) that unnecessary legal, organizational and financial barriers to access results of publicly funded research should be removed as much as possible"; moreover they welcome "open access to scientific publications as the option by default for publishing the results of publicly funded research".[4]

In September 2017, the G7 Open Science Working group, representing seven top-ranked advanced economies with the highest national wealth (United States, Japan, Germany, UK, France, Italy and Canada), expressed concerns about the speed and coherence of the transition towards open science[5]. Open and unrestricted access for a large audience to academic works is one major part of this transition. Fourteen years ago,

---

2   Bell, K., 2017. 'Predatory' open access journals as parody: Exposing the limitations of 'legitimate' academic publishing. Triple C 15 (2)

3   https://www.government.nl/documents/reports/2016/04/04/amsterdam-call-for-action-on-open-science

4   The transition towards an Open Science system http://data.consilium.europa.eu/doc/document /ST-9526-2016-INIT/en/pdf

5   https://www.rri-tools.eu/-/g7-science-ministers-meeting-2017-declaration

the Berlin Declaration on Open Access to Knowledge in the Sciences and Humanities, a milestone of the open access movement, defined open access as a "comprehensive source of human knowledge and cultural heritage that has been approved by the scientific community" and required the "active commitment of each and every individual producer of scientific knowledge and holder of cultural heritage (including) original scientific research results, raw data and metadata, source materials, digital representations of pictorial and graphical materials and scholarly multimedia material".[6]

In 2017, open access has become a significant vector of scientific communication, via an ever-increasing number of books, journals, repositories and content providers, including publishers and social media. The Bielefeld Academic Search Engine (BASE) indexes more than 116 million resources from nearly 6,000 content providers, with about 60% items freely available. Other figures confirm the dramatic growth of open access.[7] Open access is part of our reality and is here to stay, definitively. There is no way back behind the pay walls. But as a part of our reality, open access also shares and reflects existing inequalities and divides. All the authors in this volume believe in the positive qualities and promises of open access. However, they also express worries and reservations about dysfunctional developments and risks of creating even more inequalities between rich institutions, organizations and countries which can afford the costs of access scientific information as well as the transition to open access via offsetting agreements and expensive article processing charges, and all the others that cannot.

It is time for a critical assessment of open access, of its dialectics, contradictions and ambivalences. Perhaps we should just stop talking about open access and start to address the underlying, real issue of control and governance of public research. Our book is just a beginning. It is dedicated to all those who need scientific information and who share the conviction that research results are common public goods that should be disseminated and available without any restriction.

---

6   https://openaccess.mpg.de/Berlin-Declaration

7   The Imaginary Journal of Poetic Economics http://poeticeconomics.blogspot.fr/

Part One:
Global Issues

Openness as Tool for Acceleration and Measurement: Reflections on Problem Representations Underpinning Open Access and Open Science

Jutta Haider

Open access has established itself as an issue that researchers, universities, and various infrastructure providers, such as libraries and academic publishers, have to relate to. Commonly policies requiring open access are framed as expanding access to information and hence as being part of a democratization of society and knowledge production processes. However, there are also other aspects that are part of the way in which open access is commonly imagined in the various policy documents, declarations, and institutional demands that often go unnoticed. This essay wants to foreground some of these issues by asking the overarching question: "If open access and open science are the solutions, then what is the problem they are meant to solve?" The essay discusses how demands to open up access to research align with processes of control and evaluation, and are often grounded in ideas of economic growth as constant acceleration.

Introduction

In a way, the rise of open science perplexes me and I have come to ask myself: if science is opened now, then how was it closed before, by whom or by what? These are questions that defy simple answers. Still, I think, we need to ask them in order to understand more clearly the specific ways in which science and other forms of academic research are constituted by and constitutive of society today and of how central actors position themselves in relation to this institution and the knowledge it creates. The answers—in the plural—are not as simple as they might seem. Often we are presented with a simple dichotomy, introducing a fault line between an open science on one side and a closed science on the other. This builds on the idea that two homogenous blocks representing different stages of maturity face each other across a divide. The open side is typically portrayed as more advanced than the closed side, thus a 'natural' development has to occur in order for closed science to evolve into the mature, open version of science.

Yet, how is this imagined evolution to be achieved? Even a superficial reading of the types of opening strategies that are used in the debates makes it quite obvious that the aspects of science that are seen as closed are not the ones that were challenged by for instance feminist, postcolonial or post-development science studies scholars in their powerful critiques over the last decades (9). Their work showed over and over how the knowledge regimes and epistemological bases of science are deeply implicated in society's various oppressive strategies. This, as I have argued elsewhere, is not fundamentally questioned in the open access movement (7, 8), rather it is used as a scaffolding and nor is it challenged much—it seems—by proponents of open science. So, if it is not this type of closed-ness that is at stake, then what characterizes the closed science in contemporary mainstream descriptions of open science? And, importantly, for the benefit of whom is it being opened?

The problematizations underpinning open access and science

The overarching analytical question that I pose throughout this essay is: *If open access and open science are the solutions, then what is the problem they are meant to solve?* By posing this question I follow loosely in the path staked out by political scientist Carol Bacchi (1) who—drawing on Foucault's work-proposes a structured focus on problematizations in order to make visible the politics that organize assumptions and in turn the formation of issues. "Studying problematizations", Bacchi writes, "allows one to consider the relations involved in their emergence through examining how they are 'thought'" (1, p.4)—and practiced as well, I would suggest.

To do this I start with how open access was framed in the beginning and end up with its integration in a larger apparatus of open science. Other paths would be possible and are certainly relevant. However, I see one important red thread running from the establishment of the open access movement in the early 2000s to today's much larger notions of open science where open access has been surrounded by more and more concepts, all amassing into a sprawling openness apparatus to be managed, controlled and kept growing. This thread can particularly be seen in the way in which those who were the ones challenged by proponents of open access—commercial publishers—have in significant arenas come to represent and exemplify open access and open science. They do this not in the form of a counterproposal but as an appropriation (see also 6, 11).

In times past in Budapest and Berlin

As open access has established itself as an issue that researchers, universities, and various infrastructure institutions, such as libraries and academic publishers, have to relate to, it has been reshaped and—perhaps not surprisingly—mainstream actors are now amongst its most vivid promoters, if of course not the only ones.

If we travel back in time to the early days of open access, when it got its name and the movement first gained momentum, the type of open access that was staked out then was quite different from what we find today. That being said, the Budapest Open Access Initiative[1] did introduce one of today's most dominant themes, i.e. the need for acceleration:

> An old tradition and a new technology have converged to make possible an unprecedented public good. The old tradition is the willingness of scientists and scholars to publish the fruits of their research in scholarly journals without payment, for the sake of inquiry and knowledge. The new technology is the internet. The public good they make possible is the world-wide electronic distribution of the peer-reviewed journal literature and completely free and unrestricted access to it by all scientists, scholars, teachers, students, and other curious minds. Removing access barriers to this literature will accelerate research, enrich education, share the learning of the rich with the poor and the poor with the rich, make this literature as useful as it can be, and lay the foundation for uniting humanity in a common intellectual conversation and quest for knowledge. For various reasons, this kind of free and unrestricted online availability, which we will call open access, has so far been limited to small portions of the journal literature.

---

1   Budapest Open Access Initiative (2001) http://www.budapestopenaccessinitiative.org/read

In this excerpt, the problem identified is the existence of an access barrier to the scholarly literature. If we go one level higher, this access barrier creates a further problem that needs to be solved, namely that humanity is not united its "quest for knowledge" nor in a "common intellectual conversation". The solution to this problem is presented as two-fold: firstly, the "willingness of scientists and scholars" to freely share their work to advance a higher cause and secondly, removing access barriers. The latter is advanced as a technical issue. The first part—the researchers' cooperation—is seen as already in place, yet hampered by the extant access barrier. In the second part, the speed of research is highlighted as an issue to be addressed. The image invoked is that once the metaphorical floodgate that presents a barrier to the literature is opened, research will accelerate apace. The need to see an acceleration of research is closely linked to the temporal dynamic of capitalism that requires constant acceleration of economic growth (14, 17). This is a theme that will shape how open access merges into the mainstream and into open science over the following decade, however it will also diversify.

The hugely influential and signed Berlin Declaration[2] signed a couple of years later, is more technical in style and also more detailed. However some of the underlying problems that can be drawn out are in fact quite similar:

> For the first time ever, the internet now offers the chance to constitute a global and interactive representation of human knowledge, including cultural heritage and the guarantee of worldwide access. /.../
>
> In order to realize the vision of a global and accessible representation of knowledge, the future Web has to be sustainable, interactive, and transparent. Content and software tools must be openly accessible and compatible./.../
>
> Open access contributions include original scientific research results, raw data and metadata, source materials, digital representations of pictorial and graphical materials and scholarly multimedia material. /.../
>
> Our organizations are interested in the further promotion of the new open access paradigm to gain the most benefit for science and society.

Here, the problem that can be addressed with open access is the lack of an accessible global representation of human knowledge to the advantage of science and society. Also here the barrier is a technical one, and it includes

---

2   Berlin Declaration on Open Access to Knowledge in the Sciences and Humanities (2003) https://openaccess.mpg.de/Berlin-Declaration

a lack of openly accessible content, but also for instance software, metadata and so on, making archiving and accessing possible in a meaningful way. Apart from the Budapest Initiative, the Berlin Declaration directly refers to the so-called Bethesda Statement[3] as influential for its understanding of open access. Even in the last named, the problem that is indirectly seen as the one that open access should solve is that the public benefit of scientific knowledge is not being maximized. Open access, which also in the Bethesda Statement is possible since researchers already share their knowledge and ideas, is framed as a technical problem and as an issue for policy making.

The researchers, as in the Budapest Initiative, are already part of the solution. As I and others (e.g. 7, 8, 10, 15) have argued elsewhere, these early statements and others from around the same time and the visions that are bound up in them also came with a specific set of difficulties. Specifically, the way science is seen to advance along an almost inevitable positive path simply by means of publishing results in papers and where more and more progress and development flows from science. Budapest, Berlin and Bethesda all neglect the fact that scientific publications in addition to their epistemic role in which they are communicating content, also function as ways to indicate status, merit, advancement, and belonging. As such the act of publication is profoundly entangled across scholarly practices on many different levels.

For those following the discussions at the time, there were three principal arguments used to make open access an attractive proposition to researchers and policy makers. Firstly, for researchers, measurable impact in the form of increased citations was highlighted, as these can then directly feed into performance measures as the main way for structuring academic careers. Secondly, for policy makers and research funders, costs together with efficiency were advanced as gains—namely, speeding up the scientific process by making things happen faster, and cheaper. Third, the argument that the results of research funded by tax-payers should not be paid for twice, but directly be available to tax-payers without any additional costs. Here, it made sense to connect ideas of an information commons to open access (18). In order for this to make sense in turn the early focus on publications alone was not enough. Rather the entire process of doing research had to be lifted into the discussions around openness.

---

3   Bethesda Statement on Open Access Publishing (2003) http://legacy.earlham.edu/~peters/fos/bethesda.htm

London calling: accountability, efficiency and economic growth

In the years that followed the early drives, things changed. Some problematizations increasingly gained in importance while others faded and yet others diversified. Sketching this process in detail is beyond the scope of this essay. However, when in 2012 the so-called Finch group report was issued the difference that 10 years had made was quite manifest. This "Report of the Working Group on Expanding Access to Published Research Findings" (5) is an extensive document commissioned by the British government. It details various processes involved in scholarly communication and the actors involved. One section is particularly relevant for understanding what the supposed problem is that open access is meant to solve. It reads as follows (p.5):

> Improving the flows of the information and knowledge that researchers produce will promote:
>
> • enhanced transparency, openness and accountability, and public engagement with research;
>
> • closer linkages between research and innovation, with benefits for public policy and services, and for economic growth;
>
> • improved efficiency in the research process itself, through increases in the amount of information that is readily accessible, reductions in the time spent in finding it, and greater use of the latest tools and services to organise, manipulate and analyse it; and
>
> • increased returns on the investments made in research, especially the investments from public funds.

Yet, despite the authors' assertion that "These are the motivations behind the growth of the world-wide open access movement" (ibid.), many of the issues identified as being the problems that open access should solve differ from the ones identified in what could be considered the 'founding documents' of the movement. For one, the general public (or humanity) almost disappears due to a significantly more realistic understanding of the level of specialization of research fields that make much of the literature almost impermeable outside a narrow scientific community. Here, the authors call for facilitators to translate the specialized language of science into meaningful communication with the public (p.51).

In the Finch report, the problems that require open access as a solution are, if we take the list above seriously: research's insufficient transparency and accountability, the relative inefficiency and slowness of the research process, distance to innovation, insufficient returns on the investments made in research and hence inadequate contribution to public policy making and thus—ultimately—to economic growth. The language alludes to the economism of new public management, where efficiency, transparency and most prominently accountability are staples (16).

The Finch report is neither declaration nor statement and hence less sweeping in its claims. Its recommendations are detailed, often well-grounded and considerate of the complexities of scholarly work and communication. One pervading theme in the report is acceleration and growth: speeding up access, speeding up research, growth of the number of publications, of information, of innovation, and ultimately of the economy. However, in the dictum of new public management, economic and other kinds of growth or increased returns on investment only exist when they are measured (ibid.). Hence accountability and transparency, typically translated into often numeric indicators make their entry. Publishers are seen as playing a key part in this. Curiously the universities' own repositories are scarcely mentioned in the path towards open access that is sketched in the Finch Report. Open access is turned into gold open access and gold open access is equated with the commercial publishers' version of it—pay to publish rather than for instance open access journals that are financed through grants, memberships, universities, or public funding as established for national journals in South America.

In the Finch report a few staple issues are engaged and these still re-appear and shape much of the mainstream discourse on open access and now also open science. Open access, at least from the vantage point of policy makers and research funders, is now primarily a business model for managing relations between public funders and private enterprise, tied to the scientific community through performance indicators with the ultimate aim to increase efficiency, enable (commercial) innovation and economic growth. It is now also just a puzzle piece in a larger idea of open science, in which the entire research process needs to be opened up according to a very specific understanding about its previous closure.

The New Berlin: cash flows and evaluations

This business model approach becomes very palpable in the way in which the libraries' role is described. In the OA2020 Initiative for the large scale transition to open access—a follow-up to the Berlin Declaration—libraries get a prominent role and in the roadmap to implementation they get their own headline "The transition begins with libraries" right after the introduction (13). Yet, the role of libraries is quite specific:

> As libraries are the organizers of the cash flows in the subscription system, they are the ones who must show leadership in grasping that their acquisition budgets need to be liberated and reinvested in open access publishing services. Libraries are also predestined to be the organizers of the cash flows in an open access publishing system, because they have the skills, the experience with publishers and the staffing to take care of the necessary administration. Their implicit challenge is that they must evolve their roles, responsibilities, profiles and workflows.

Considering that a considerable part of the transition once did begin with libraries and librarians, it is interesting that what counts here is not their experience with open access or even their role in scholarly communication, collection building and in preserving and making accessible the records of science, but their experience with publishers and budgets. After all, libraries and librarians were one of the earliest driving forces behind open access and in the course of their advocacy they have already considerably evolved their "roles, responsibilities, profiles and workflows" as they lobbied university management, funders, researchers, promoted open access to the media and the public, developed and maintained institutional repositories, directories of open access journals, held lectures and seminars, organized conferences about open access and much more.

Furthermore, leaving aside the constrained position libraries are in regarding control over their assigned budgets, what is fascinating here is that this is the only role libraries are seen to have for open access in this roadmap: dealing with the budget and the costs of open access as well as administrative duties related to redirecting the cash flow from subscriptions to open access publications, presumably author fees. In most cases the recipients of this re-organized cash flow will be the same publishers that previously charged subscription fees (see also 11, 12).

The OA2020 initiative is interesting also in a different way. In the Finch report we could see the language of new public management's audit culture shine through: open access should usher in improved efficiency, increased return on investments, accountability, and innovation. The OA2020 initiative sees itself as "one element of a more profound evolution of the academic publishing system that will lead to major improvements in scholarly communication and research evaluation" (13). An exact discussion of how open access should improve research evaluation is beyond the scope of the initiative, yet statements like these tie open access closer to a form of administrative enclosure through which research policy actors express and enact control. Open access has been turned into an indicator, one amongst several, to motivate researchers and to assess their performance.

Brussels taking interest: acceleration galore and disruption as the new normal

The new reality of open access as an indicator becomes apparent in the final document that I want to discuss, the European Commission's 2016 publication "Open innovation, open science, open to the world—a vision for Europe" (3). Here it becomes clear just how many different concepts have been latched onto open access. Open access has by now become just a small wheel in a type of *deus ex machina*, in which a specific kind of ideologically confined, technical openness becomes part of an imagined transformative system change that is almost entirely impregnated in the language—some would say jargon—of economic necessity, commercial interests and technological determinism.

The document consists of four parts. The first three are reflected in the title and the last contains a selection of tone-setting speeches by Carlos Moedas, European Commissioner. Other fundamental concepts discussed are open data and citizen science. The way in which open science is couched between a section on Open Innovation and a part called Open to World is symptomatic of the conflicting framing of the issue, a conflict between a notion of universal science collaboratively produced for the common good and "capitalism's speed imperative" that within the structures of academia is often enacted as "competitiveness talk" (17, p.35 and 95seq.). Open science itself has to become part of a competition:

> Ensuring Europe is at the forefront of open science means promoting open access to scientific data and publications /.../. (p.7)

Throughout the text, and certainly in the speeches, the most eminent reasons for implementing open science and open innovation are framed in terms of competitions. Allusions are made to a race in science in which Europe is lagging behind or at risk of falling behind and which open science should help Europe win. Acceleration, speeding-up, a theme that was already established in the Budapest Open Access Declaration, fifteen years earlier, is a defining feature of this vision for open science in Europe. It is translated into winning a race, topping a league table, winning Nobel prizes, taking on the USA, and importantly transforming scientific results into commercial output, i.e. innovation, which should put European industry ahead in various ways. Crucially, open science is to be helped by open innovation "to connect and exploit the results of open science and facilitate the faster translation of discoveries into societal use and economic value" (3, p.15) and so on.

The document is packed with metaphors alluding to speed, races, rankings, competition, the need to catch up—in short today's (closed) science is too slow, openness means acceleration and acceleration means winning the all-defining, all-encompassing race progress is seen to be. Superficially, this stands in some contrast to an "open to the world" attitude which is also present in the text and where global challenges (e.g. Zika, Ebola, food, water, health and energy) must be solved across national borders. Yet, even collaboration is cast in a terminology of competition: "global competition for talent" (p.60), "the rapid rise of China" (p.60), although "Europe has been able to maintain its lead in terms of highly cited papers" (p.61), yet the US is, at least according to certain measurements, better at collaborating with Asian countries, a problem that means "All available instruments are put to use to maximize the impact of international cooperation on research and innovation"(p.64). In short, open science is a means to win a race, even when it is about working together. The market-place rhetoric of competition is also reflected in the general framing of open science in the section's introduction, in which science and open science are likened to businesses:

> Open science is as important and disruptive a shift as e-commerce has been for retail. Just like e-commerce, it affects the whole 'business cycle' of doing science and research—from the selection of research subjects, to the carrying out of research and to its use and re-use—as well as all the actors and actions involved up front (e.g. universities) or down the line (e.g. publishers). (3, p. 33)

The image of disruption of retail is here presumably intended as a positive signal for change, yet it also gives way to images, I suggest, of a concentration of power and wealth in the hands of fewer and fewer internet companies, of abandoned city centers, of labor market deregulation and increasingly worsened working conditions for workers in the so-called gig-economy, and exploitation of free labor and automation of work.

Closed is open and open is closed: openness as a performance indicator

In the early days of the open access movement and to a degree also in the Finch report, scholars were framed as part of the solution to the respective problem that was identified. In the latest version of open science, their role is imagined differently. In fact the tables seem to have turned. Here scholars are part of the problem while those entities such as publishers that once were perceived as barriers have come to signify openness.

In a figure in the report (3, p.15) open science is illustrated by various digital services from discovery, analysis, writing, publication, to outreach and assessment. The services and tools are presented in four rows, whereas three list services by Elsevier, Springer, and Google. The last row lists services by Wikimedia. Publishers and the biggest internet company of them all, Google, have here come to illustrate and signify open science. The roles of researchers have become a lot less obvious. Theirs is a role that needs to change, adapt, transform, be surveilled, incentivized and most of all improved.

For this, open access and open science are turned into performance indicators to be used for evaluating individual researchers. "One incentive is to integrate open access in the evaluation of a researcher's career" (3, p.50) a statement reads on the same page as the illustration described above. The notion of competition and acceleration that underpins much of the contemporary mainstream discourse of science (17) and which is often expressed in terms of various indicators (2) is now applied directly to open science. This discourse is framed as a transformation, a disruption of the system, thus further solidifying the imaginary of open science as part of the never-ending acceleration and economization of society.

Openness: acceleration, measuring and economization

In this short essay, I argue that a significant part of how open access and increasingly open science is imagined today and specifically of how mainstream actors, such as funders and national or supra-national policymakers present it, is part of a drive towards a further economization and privatization of a vast and diverse field of publicly funded knowledge production. This is not the case, for all of open science, of course, as evident in numerous alternative scholar-driven projects, often in the humanities, the arts and related fields, which are about a different kind of change.

Yet, the version that has traction, and policy making and funding behind it, needs to be understood, I suggest, within larger tendencies of capitalist dynamics of constant growth, acceleration and value accumulation framed in simple trajectorial narratives (4, p.134–136) in which progress largely means following a linear path towards increased efficiency while chasing a relentlessly delayed future. To get to the bottom of this— in order to enable alternative narratives and to attend to diverse ways in which actual research practices are reshaped as a response—is a large project. Here I could only sketch some preliminary lines of analysis while zooming in on a few selected instances of how the aspired openness and the imagined closed-ness of science are given meaning in relation to each other on a macro-level.

References

(1) Bacchi, C. (2012). Why study problematizations? Making politics visible. *Open Journal of Political Science*, 2(1), 1–8.

(2) De Rijcke S., P.F. Wouters, A.D. Rushforth, T.P. Franssen, and B. Hammarfelt (2016). Evaluation practices and effects of indicator use: A literature review. *Research Evaluation*, 25(2), 161–169.

(3) European Commission (2016). *Open Innovation, Open Science, Open to the World: A Vision for Europe*. Brussels: Directorate-General for Research and Innovation.

(4) Felt, U. (2016). Of Timescapes and Knowledgescapes. Retiming Research and Higher Education. In P. Scott, J. Gallacher, and G. Parry (Eds.), *New Languages and Landscapes of Higher Education*, pp. 129–148. Oxford: Oxford University Press.

(5) Finch Group (2012). *Accessibility, Sustainability, Excellence: How to Expand Access to Research Publications*. The Association of Commonwealth Universities, London.

(6) Guédon. J-C. (2017). *Open access: Toward the Internet of the Mind*. Unpublished paper, accessible http://www.budapestopenaccessinitiative.org/boai15/Untitleddocument.docx

(7) Haider, J. (2007). Of the rich and the poor and other curious minds: On open access and "development." *Aslib Proceedings: New Information Perspectives*, 59(4–5), 449–461.

(8) Haider, J. (2012). Open access hinter verschlossenen Türen oder wie sich open access im und mit dem Entwicklungsdiskurs arrangiert. In U. Herb (Ed.), *Open Initiatives: Offenheit in der digitalen Welt und Wissenschaft*, pp. 65–84. Saarbrücken: universaar.

(9) Harding, S. (2006). *Science and Social Inequality: Feminist and Postcolonial Issues*. Champaign, IL: University of Illinois Press.

(10) Herb, U. (2010). Sociological implications of scientific publishing: Open access, science, society, democracy, and the digital divide. *First Monday*, 15(2), 1 February 2010.

(11) Herb, U. (2017). Open access zwischen Revolution und Goldesel. In *Information. Wissenschaft & Praxis*, 68(1), 1–10.

(12) Lawson, S., J. Gray, and M. Mauri (2016). Opening the black box of scholarly communication funding: A public data infrastructure for financial flows in academic publishing. *Open Library of Humanities*, 2(1), p.e10.

(13) OA2020 (2015). *Initiative for the Large Scale Transition to Open Access, Roadmap*. Munich: Max Planck Digital Library.

(14)   Rosa, H. (2015). *Social Acceleration: A New Theory of Modernity*. New York: Columbia University Press.

(15)   Schöpfel, J. (2015). Open access—the rise and fall of a Community-Driven model of scientific communication. *Learned Publishing 28* (4), 321–25.

(16)   Shore, C., and S. Wright (2015). Governing by numbers: Audit culture, rankings and the new world order. *Social Anthropology* 23(1), 22–28.

(17)   Vostal, F. (2016). *Accelerating Academia: The Changing Structure of Academic Time*. Houndmills, Basingstoke: Palgrave Macmillan.

(18)   Willinsky, J. (2005). *The Access Principle: The Case for Open Access to Research and Scholarship*. Cambridge MA: The MIT Press.

Open Access, a New Kind of Emerging Knowledge Regime?

Elena Šimukovič

Open access as a programmatic name for a new mode of dissemination of schol-arly publications has been around since the turn of this millennium. However, a considerable increase in calls for a more rapid transition from journal sub-scription to "full" open access system can be observed in recent years. By looking at some of the beginning aspirations of the open access movement as well as proposed disruption scenarios, this contribution aims at discussing some of the less visible aspects of current debates. The ultimate aim is to embed this new understanding of the open access system within the conceptual framework of an emerging "knowledge regime".

Introduction

For quite some time now, academic libraries, research and funding orga-nizations and the individual actors behind them have been noticeably concerned with one topic. The long-cherished quest for the universal acces-sibility of humanity's (scientific) knowledge seems to have come within reach. The conventional subscription model for publishing scholarly jour-nals is supposed to undertake a radical shift and to transition into a full open access system in which everyone is able to read about research results without having to overcome the "pay wall". As much research takes place in higher education and other research institutions that are supported with public funds, it is hard to disagree with the goal of making the fruits of this research available to the public as well.

However, the concept and practice of open access has a long history of developing various technical, legal and financial implementation models, which makes speaking of one homogeneous "open access movement" mostly a matter of symbolic unity rather than analytic coherence. The considerable amount of attention to open access in the science policy arena indicates a new stage in its evolution. Looking retrospectively and prospectively at developments in open access can help us predict the possible evolution of academic publishing and scholarly communication.

Full Open Access as a long-cherished wish

The transition to a full open access system, one in which scholarly publications are universally accessible on the internet and not only to institutional or individual subscribers, seems to be an intrinsic desire in the history of open access. From its inception at the turn of this millennium it was a central idea that open access would come to replace other modes of granting access to scholarly literature. One of the illustrative examples can be found in the Budapest Open Access Initiative (BOAI), which was published in February 2002 and is widely regarded to be the first venue that used the term "open access" and articulated a corresponding definition. BOAI proposed not only a vision of an "unprecedented public good"—one in which a new internet technology and an old tradition of sharing fruits of research will converge—but also encouraged experimentation with different implementation strategies "to make the transition from the present methods of dissemination to open access". To mark the tenth anniversary of the original declaration, BOAI reaffirmed its aspiration for a widespread adoption of open access and added a spatio-temporal dimension for it to "become the default method for distributing new peer-reviewed research in every field and country" within the next ten years (4).

Various research and funding organizations have also supported the idea of full open access in the universe of scholarly communication. Major associations such as Science Europe, the European University Association (EUA) and the Global Research Council (GRC) have endorsed action plans and issued their own transition principles, urging all stakeholders to replace "the present subscription system with other publication models whilst redirecting and reorganizing the current resources accordingly" (12). Along with numerous institutional open access policies such demands mainly tackled research publications that result from endeavors supported by public funds. Somewhat unsurprisingly, the parties involved mostly agree on the basic principle behind this rationale—to make publicly funded research results freely accessible to the public. However, points of contention show up when it comes to the ways to translate this apparently simple principle into practice. This includes the full "color spectrum" of so-called green, gold, diamond, platinum and other species of open access models as well as pilot agreements with academic publishers at an institutional or a national level (2).

However defined in its practical terms, "open access" has embodied a pivotal element in many science policy related debates in the past few years and has served as a powerful heading to bring different actors together and to mobilize resources. Open access has been declared a strategic goal in several, mostly European countries such as Austria, the Netherlands or Sweden that now aim for 100% of unrestricted access to their research results by a specified date (3). Moreover, the issues of "openness" in scholarly communication have received amplified attention with the advent of "open science." This is a fellow concept to open access that incorporates free access not only to the final publications, but also to the underlying data or methods used. These issues were put on the official agenda of the Dutch Presidency of the Council of the European Union in the first half of 2016. In this context, the "Amsterdam Call for Action on Open Science" was released in April 2016 (1), following the launch of the "OA2020" campaign "for the large-scale transition to open access" by the Max Planck Society in March 2016 (11). Both initiatives joined together in formulating "a clear pan-European target: from 2020 all new publications are available through open access from the date of publication" (4). The following section will thus take a closer look at proposed transition paths to full open access and the possible implications thereof.

Time for disruption? Designing an Open Access transition

The foundation for the launch of the OA2020 campaign was laid in a "white paper" by the Max Planck Digital Library (MPDL) in April 2015 (11). Its core message, namely that there is "already enough money in the system" and a large-scale transformation from subscription to open access publishing would be "possible without added expense", was at the heart of many subsequent debates. For instance, in a joint statement in October 2015 EU Commissioner Moedas and Dutch Secretary of State

Dekker called upon academic publishers "to adapt their business models to new realities" (6). In the 21st century, they continue, academic publishing has to move to open access and only fair and fully transparent business models will be accepted. In a very similar vein, the League of European Research Universities (LERU) urged publishers "to enter a brave new [open access] world" (10).

Interestingly enough, the line of reasoning in the OA2020 initiative repeatedly reads as stating an inevitable development: "One doesn't need a crystal ball to predict that the subscription system will come to a natural end, sooner or later. An unmanaged end would probably be more unpredictably disruptive, and could create damage to the meaningful core activities of the current publishing infrastructure."[1] Best parts of the new and the old worlds are thus supposed to be combined, with the "disruptive element" for an orderly transformation "directed at the financial streams and the business models only". For this to be accomplished in practice, the authors of the paper promote a strategy of shifting the payment streams from subscribing to academic journals (as it has been the case so far) towards publishing all scientific articles in fee-based open access journals, whereas conventional subscription journals are expected to eventually "flip" to full open access. This scenario means that academic libraries across the globe are requested to repurpose their acquisition budgets and to cover author-side publishing fees (Article Processing Charges, APCs) for members of their research institutions instead of subscription or licensing fees (11).

Despite the wide uptake of the initiative[2], there have been less enthusiastic reactions to the proposed transition path that have stressed the drawbacks of an exclusive APC-focused open access model and unintended consequences that such a move could have. One of the notable examples includes a joint statement by the Confederation of Open Access Repositories (COAR) and United Nations Educational, Scientific and Cultural Organization (UNESCO) (5). With their main concern being the particular kind of open access that is promoted in this campaign, they voice misgivings with respect to researchers located at institutions with tight budgets or in less affluent countries: "Authors will be unable to publish once limited funds have been exhausted. Such a system will need to support researchers who cannot pay APCs—to avoid further skewing a

---

1   FAQ, OA2020, accessed February 7, 2017. https://oa2020.org/faq/

2   As of February 7, 2017, 73 scholarly organisations have officially signed the Expression of Interest.

scholarly publishing system that is already biased against the research undertaken in certain disciplines and countries". Secondly, they assume that a flip to an APC system would lead to further concentration in the international publishing industry that is already dominated by the biggest publishers: "A mere shift towards the pay-to-publish model will institutionalize the influence of these companies, and discourage new entrants and models other than APC models". Thirdly, new open access models are required to build in mechanisms that ensure cost reductions: "Simply shifting payments to support APCs may lead to higher systemic costs, curb innovation, and inhibit the scholarly community's ability to take advantage of new models and tools".

Following UNESCO's World Science Report, both organizations further emphasize the critical role that scientific knowledge plays in socio-economic welfare and the global economy. No region or nation can remain a simple "user" of knowledge but must also become a "creator" of new knowledge. Thus, they warn that a "large-scale continental shift towards a pay to publish model in Europe may have significant unintended consequences for both Europe and elsewhere by impeding global participation in the system" and ask governments and the research community to look for a variety of approaches for a healthier and more innovative ecosystem (5).

Thinking in terms of emerging knowledge regimes

Since multiple efforts to accelerate progress towards open access as a default mode of academic publishing have built up momentum and even manifested themselves in national and international policy goals, it might be helpful to consider recent initiatives from a broader conceptual perspective. Particularly, looking through the lens of Science and Technology Studies (STS) and related research domains can offer a hint at potentially neglected or otherwise invisible aspects in current debates. The notion of emerging knowledge regimes will thus serve as an illustrative example[3].

Building on "technopolitical regimes" developed by Gabrielle Hecht in her work on the role of nuclear technologies in shaping the national identity in France, the "regime" metaphor offers three basic ideas to be considered. First, its political parlance and the capacity "to refer at once

---

3   For bringing other theoretical concepts into play see also (13).

to the people who govern, to their ideologies, and to the various means through which they exert power", including "their guiding myths and ideologies, the artifacts they produce, and the technopolitics they pursue" (8). Second, to point at the prescription of certain policies and practices as well as broader visions of sociopolitical order. And finally, to emphasize the contested nature of power in which different regimes always have "to contend with varying forms of dissent or resistance, both from outside and from within the institutions they governed".

In the context of large-scale open access transition initiatives as described in the previous section, the concept of "knowledge regime" then aims both to evoke similarity with political regimes and to convey the difference that knowledge makes. It is important to bring again to the fore at this point, that various models to "opening" access to this knowledge are available, of which so-called green and gold open access are the primary examples. Although both models were proposed initially as complementary strategies[4], current open access discussions often gravitate around an either-or dichotomy or even put the two in opposition. For instance, the OA2020 initiative committed itself from the onset to reject the green model, which involves delivering access to scholarly publications by means of depositing author manuscripts in online repositories. The authors give the following reasons for this standpoint: "Although green approaches have been around for the past 20 years, they have not led to any progress regarding the subscription system. On the contrary, subscription spending has steadily increased during this period. There is no indication of any delegitimization of the prevailing distribution and financing conditions by means of the green route to open access. The green approach has not proven to be an effective tool to overcome the publishers' market power, to bring down prices and to establish open access on the grand scale"[5]. Thus, the career of and the relation between the green and gold siblings seem to have experienced their own evolution in terms of the number of respective supporters and expectations on their potential impacts.

To push the argument even further, it can be said that different targets may be pursued by giving preference to one or another model. For instance, if one would strive for the most cost-efficient option, green open access might be the likely answer as publication manuscripts can

---

4   BOAI, "Read the Budapest Open Access Initiative," 2002; the labels "green" and "gold" were proposed somewhat later, though.

5   FAQ, OA2020, accessed February 7, 2017. https://oa2020.org/faq/

be deposited to (often already existing) institutional or subject repositories at nearly zero cost. If one in turn wishes to foster alternative publishing venues in first place, community-driven gold open access and novel funding models such as the Open Library of Humanities[6] might help to level the playing field. However, massive investments in bulk prepayments to ransom individual articles in so-called "hybrid" subscription journals might risk perpetuating the status quo in power relations and the price spiral in the academic publishing system. Furthermore, should the argument of making publicly funded research available to the public be taken seriously, would the regular citizens pay as much attention as their fellow scientists to the title of the journal in which an article appeared? Or would they rather, for instance, make use of local newspapers that discuss the relevance of specific results and link to respective repositories from which a full-text copy can be downloaded?

Although it is a highly speculative undertaking with no definitive answers—given the evolving state of affairs as well as the complexity of the issues at stake—engaging with underlying assumptions in such transition scenarios could potentially shed light on some of the potential drawbacks in this new kind of knowledge regime. Moreover, to regard certain phenomena as "emergent", that is, to put emphasis "not on the unfolding of something already in being but on the outspringing of something that has hitherto not been in being,"[7] suggests that emergencies can also be perceived as threatening because they come along with an inherent portion of uncertainties and insecurities. Thus, a disruption of scholarly publishing towards full open access will inevitably cause a pinch of discomfort as it entails a "side effect" to destabilize current workings of the science system and to redraw at least some of its boundaries.

Concluding remarks

It has become visible from many initiatives and related debates that open access is "here to stay". While taking into account the pros and cons of different roads towards a desirable future, different actors might go in different directions according to their preferences or available resources.

---

6   "About," Open Library of Humanities, accessed February 7, 2017. https://www.openlibhums. org/site/about/

7   Morgan, 1927: 112, cited in (9): 66.

Bearing this kind of multiplicity in mind, or in some cases even diverging developments, it seems more appropriate to speak of different and sometimes opposed branches rather than *the* open access movement.

As several parties noted in their comments and position statements, relying on one particular academic publishing model—whatever it would be—cannot be expected to suit researchers in all fields and life course situations as a one-size-fits-all approach. Mainstreaming advocacy campaigns would thus need to acknowledge the plurality of options and allow for "deviant" directions departing from one-way, binding roadmaps. Questioning the dominance of gold (or rather hybrid) open access models and heavy focus on scientific journals published by major commercial publishers should be open to scrutiny as much as other alternatives allowing for variety of "legitimate" research profiles and publishing practices. Although current large-scale initiatives are surrounded by a considerable amount of noise and hype, their present orientation might lead to a fundamental fallacy of pushing for an immediate yet pricy "accessibility" of scientific articles but do little to alter established copyright transfer practices or issues in re-usability of research results[8]. Instead of reaching this unsatifying impasse, the proposed (r)evolution in scholarly publishing could benefit from an effort to avoid merely relocating the "entrance fee" to the scholarly publishing world from a "pay-to-read" towards a "pay-to-say" threshold. Being responsive to the intricate relationships and difficult entanglements between science, technology, society and policy realms is not a straightforward task. However, doing so is essential to fostering the long-cherished quest not only for universal access, but also participation in producing and sharing knowledge across the globe.

---

8 Some readers might recall discussions on "gratis" versus "libre" Open Access at this point. See (7) for instance.

References

(1) The Netherlands EU Presidency (2016). *Amsterdam Call for Action on Open Science*. http://english.
     eu2016.nl/documents/reports/2016/04/04/amsterdam-call-for-action-on-open-science.

(2) Association of Universities in the Netherlands (VSNU) (2017). *Greater impact with open access*.
     http://vsnu.nl/more-impact-with-open-access/index.html.

(3) Bauer, B., et al. (2015). *Recommendations for the Transition to Open Access in Austria*. http://dx.doi.
     org/10.5281/zenodo.34079.

(4) Budapest Open Access Initiative (2012). *Ten years on from the Budapest Open Access Initiative:
     setting the default to open. BOAI10 Recommendations*. http://www.budapestopenaccessinitiative.org/
     boai-10-recommendations.

(5) COAR, and UNESCO (2016). *Joint COAR-UNESCO Statement on Open Access*. http://www.
     unesco.org/new/fileadmin/MULTIMEDIA/HQ/CI/CI/pdf/news/coar_unesco_oa_statement.pdf.

(6) European Commission (2015). *Commissioner Moedas and Secretary of State Dekker
     call on scientific publishers to adapt their business models to new realities*. News release.
     October 12, 2015. http://ec.europa.eu/commission/2014-2019/moedas/announcements/
     commissioner-moedas-and-secretary-state-dekker-call-scientific-publishers-adapt-their-business_en.

(7) Harnad, S. (2011). *Gratis Open Access Vs. Libre Open Access*. http://openaccess.eprints.org/index.php?/
     archives/862-Gratis-Open-Access-Vs.-Libre-Open-Access.html.

(8) Hecht, G. (2001). Technology, Politics, and National Identity in France. In Allen, M.T., and G.
     Hecht (Ed.), *Technologies of power: Essays in honor of Thomas Parke Hughes and Agatha Chipley Hughes*,
     pp. 253–294. Cambridge, MA: The MIT Press.

(9) Hodgson, G.M. (2000). The Concept of Emergence in Social Sciences: Its History and Importance.
     Emergence, 2(4), 65–77.

(10) League of European Research Universities (LERU) (2016). *The academic world urges publishers to
     enter a brave new world*. News release. January 27, 2016. http://www.leru.org/index.php/public/news/
     the-academic-world-urges-publishers-to-enter-a-brave-new-world/.

(11) Schimmer, R., K.K. Geschuhn, and A. Vogler (2015). *Disrupting the subscription journals' business
     model for the necessary large-scale transformation to open access. A Max Planck Digital Library Open
     Access Policy White Paper*. http://hdl.handle.net/11858/00-001M-0000-0026-C274-7.

(12) Science Europe (2013). *Principles for the Transition to Open Access to Research Publications*. http://
     www.scienceeurope.org/uploads/PublicDocumentsAndSpeeches/SE_OA_Pos_Statement.pdf.

(13)   Šimukovič, E. (2016). *Of hopes, villains and Trojan horses—Open Access academic publishing and its battlefields*. Doctoral research proposal, Department of Science and Technology Studies, University of Vienna. http://hdl.handle.net/10760/29265.

Open/Access: Negotiations Between Openness and Access to Research

Samuel A. Moore [1]

Open access (OA) is a contested term with a complicated history and a variety of understandings. This rich history is routinely ignored by institutional, funder and governmental policies that instead enclose the concept and promote narrow approaches to open access. This paper presents a genealogy of the term open access, focusing on the separate histories that emphasise openness and reusability on the one hand, as borrowed from the open source software and free culture movements, and accessibility on the other hand, as represented by proponents of institutional and subject repositories. From analysing its historical underpinnings and subsequent development, I argue that open access is best conceived as a boundary object rather than a policy object because boundary objects lose their use-value when "enclosed" at a general level, but should instead be treated as a community-led, grassroots endeavour.

Introduction

The concept of open access resonates differently between communities of practice. Broadly speaking, it refers to the removal of price and permission restrictions to scholarly research. Open access research is free to read and use by anyone with access to a stable internet connection. This definition is

---

1 The author is also a part-time employee of the open-access publisher Ubiquity Press. They had no role in the research, preparation or writing of the manuscript.

generally consistent across communities, although some insist on a specific, permission-free approach to open access licensing, while others specifically discourage the use of liberal licensing without limits to reuse. However, it is the motivations for open access and routes to it that differ most substantially between communities, which will be the focus of this chapter[2].

The development of open access reveals a number of different lineages, from the formalising of pre-existing preprint cultures via subject repositories and the emergence of institutional repositories, to the free culture and open source software movements. These separate lineages do not make for a consistent set of values associated with open access, especially against the backdrop of the range of disciplinary publishing cultures and working practices. Throughout the chapter I argue that these numerous motivations and values mean that open access should not be approached as a unified whole. One should therefore not think of open access as a thing-in-itself; rather, it should be seen as a process of understanding, engaging and experimenting with the ways in which research is presented and disseminated. Open access should therefore be considered and fostered as a community-led initiative; and funders, institutions and governments should be deferential to this in their approach to policymaking.

The chapter takes a genealogical approach to the history of open access through an investigation into the separate and often conflicting discourses that have shaped ideas around openness and access to scholarship. The genealogical approach is associated with the work of Foucault (inspired by Nietzsche's method in Genealogy of Morals) and involves cultivating "the details and accidents" involved in the origin and development of a concept (5, 80)—in Foucault's case the concept of "discipline", for example. By analysing the tactics used in employing the concept, a genealogy will reveal the power that shapes and governs a particular discourse. It is, in Foucault's words, a way of revealing the "history of the present" (6, 31). For this reason, my aim here is not to provide a conclusive or exhaustive account of the history of open access, rather to illustrate how some of the different histories of open access have resulted in the landscape that exists now.

The chapter discusses two distinct lineages of open access that have in various ways converged in contemporary understandings of the term.

---

2   This is an edited and condensed version of an article published by RFSIC (*Revue Française des Sciences de l'Information et de la Communication*) under an CC-BY-NC-SA license: Moore, S. A. (2017). A genealogy of open access: negotiations between openness and access to research. *Revue française des sciences de l'information et de la communication*, Vol. 11. http://journals.openedition. org/rfsic/3220 ; DOI : 10.4000/rfsic.3220

Analysing both discursive and non-discursive articulations of open access, I illustrate how there are some formulations that derive from attempts to provide cost-free access to research works, such as those associated with institutional repositories, subject repositories or early open access journals on the web. On the other hand, there are approaches that derive more from open source software, such as those associated with new journals in the biological sciences or those advocating libre Creative Commons licences that permit reuse of work in accordance with certain contexts. These approaches emphasise the open nature of research: it should be reusable and re-mixable, all for commercial purposes, in a similar way that open source software is. Further still, I illustrate the unique motivations for open access, and routes to it, that can be found within these two distinct lineages, such as the desire to reduce subscription prices or those associated with a particular political position, be it market-based or progressive.

I employ a genealogical approach to understand the many ways in which open access came into being, in order to illustrate the term as multiple, processual and responsive to a range of motivations. This conception naturally lends itself to what Star and Griesemer term a "boundary object"—a concept that has a specific understanding in a local community of practice but is rigid enough to maintain its definition across communities too (Star and Griesemer 1989). As such, boundary objects can be approached and understood at a general level, between communities, but they also permit experimentation and community ownership of the object at hand. Open access, as I show, has been successful precisely because it resonates across communities, but the history of open access will also illustrate that it has specific meanings in individual circumstances.

Clearly, this genealogy will be incomplete and somewhat oversimplified, but it will highlight the community-specific nature of open access and the need to not enclose it according to a rigid, sweeping understanding of the term. It will also highlight the hegemonic struggles involved in the development of open access and the need to ensure that the development of open access is not solely driven in accordance with the interests of more dominant groups. I will begin by detailing what I argue are the two separate approaches to open access that eventually converged in the mid-2000s, specifically those that emphasise "openness" versus those prioritising access to research.

Openness

In a general sense, openness refers to the degree to which a thing or action is freely accessible. It implies freedom: the extent to which a particular action, resource or concept is free to perform, access or use. It also implies transparency, where, for example, governments share their accounts under the label of open government data or simply where one speaks frankly and does not self-censor. But openness appears to be a term with multiple understandings and no fixed definition. It is cited by governments, startups and organisations as integral to their "philosophy", often without further explanation of the term. As Nathaniel Tkacz argues: "Somewhat ironically, once something is labelled open, it seems that no more description is needed…[O]penness is the answer to everything and what we all agree upon" (16, 37).

Tkacz traces a line from Karl Popper's conservative discourse *The Open Society and Its Enemies* to the open source movement of the 1980s and beyond, focusing in particular on how this continues to influence contemporary understandings of openness. The open source method of software development is, for Tkacz, representative of the kind of decentralised, competitive approach to government developed by early theorists of neoliberalism such as Popper and Hayak. Coupled with its specifically pro-business outlook, it is clear to see a link between open source and neoliberalism, especially as the Linux operating system (often hailed as the crowning achievement of open source software) is used by multinational corporations everywhere, including Google and Amazon.

If we are to accept Tkacz's account of openness-as-neoliberalism and look closely at the projects that bear its name, we would surely find that open access itself bears the same neoliberal hallmarks. Certainly, many aspects of open access were influenced by open source software, particularly the use of open access licences in place of traditional copyright. For Tkacz, open projects display varying degrees of "transparency, collaboration, competition and participation"—all of these qualities are observable in the usage of the Creative Commons CC BY licence, which permits readers the right to freely read, share and reuse published research (for commercial purposes) without requiring permission from the copyright holder.

As is well known, CC BY is widely used within open access scientific publishing. It is the dominant licence of the new "megajournal" business model that publishes research irrespective of its perceived importance. Article Processing Charges are closely associated with this highly profitable approach to open access and many millions are spent by funders and universities on them each year (see e.g., Lawson 2015). Clearly open access has been opened up to competition within the free market, despite one of the primary motivators for open access being an objection to the profiteering practices of commercial publishers. Further still, in favouring CC

BY over other more restrictive licenses, published articles are open for reuse by commercial entities, further illustrating the relationship between certain articulations of open access and the free market (and neoliberalism more generally) in the way Tkacz describes.

When understood through the history of open source software, then, it is clear that some understandings of openness promote a neoliberal vision along the lines described above. In many respects, openness is pragmatic, business-friendly, competitive and non-centralised; it has been easily embraced and subsumed by capitalism in the same way as many instances of open source software have. However, just because openness (and open access specifically) can be "neoliberalised", it would be an overgeneralisation to assert that all instances of open projects derive from the intellectual project of neoliberalism, as I will explain forthwith.

The antagonisms of openness

The problem with generalising out from the politics of some projects that operate under the banner of "open" to all of them is that it treats the political in general as a category that has already been decided upon, rather than a decision made, as Chantal Mouffe illustrates, in an "undecidable terrain" (14, 17). "Neoliberal" is not a political category that can be indiscriminately applied to all forms of openness but something operating in a specific context and under certain conditions (or "enclosures"). Tkacz himself recognises this, stating:

"Rather than using the open to look forward, there is a pressing need to look more closely at the specific projects that operate under its name—at their details, emergent relations, consistencies, modes of organising and stabilizing, points of difference, and forms of exclusion and inclusion [...]" (16, 38).

It is these "details, emergent relations, modes of organising and stabilizing, points of difference, and forms of exclusion and inclusion" that contribute to a project's politics. These enclosures need to be made and constantly reassessed, rather than decided upon in advance as a homogenous category or structure.

Gary Hall makes this point about open access specifically: "to argue that open access is political in this explicit, a priori way, would be to give the impression that it is so simply because it conforms to some already established and easily recognized criteria of what it is to be political" (9, 35–36). Certainly, some examples of openness (and open access) do conform to the rhetoric of the market and competitive, individualised approaches scholarship. But other examples of openness are more progressive, seeking instead to organise in a way that tackles a specific problem in a given context. The status of openness as "political" in any form (be it progressive or reactionary) is not something that can be decided upon in advance.

It is difficult, then, to speak of openness as a thing-in-itself without modifying or enclosing it in some way. Openness of course broadly refers to the gifting of the outputs of one's creative or intellectual endeavours in accordance with certain conditions. It is the choices made around how this is done, what enclosures are made and how projects are organised that make up their politics. For example, Dymitri Kleiner's peer-production licence is a form of open licence that aims to foster the creation of a commons so that "independent communities of peers can be materially sustained and can resist the encroachments of capitalism" (11, 12). To achieve this, Kleiner modified the Creative Commons Sharealike (CC BY-SA) licence to prohibit the reuse of works by for-profit corporations. For-profits are able to reuse licensed works but only after paying a fee to their creators. This encourages a different kind of commons, based on sharing via a copyleft clause (one which the author terms "copyfarleft"), but one which confronts what Kleiner sees as an "unfree society that requires consumer goods to capture profits" (11, 28).

Kleiner's peer-production licence represents an attempt to use free culture to promote a specific kind of politics. This involves a kind of antagonism or enclosure, i.e., an active choice as to the way things should be in a particular context. Antagonisms are the foundation of the political sphere; they represent disagreements or conflicts over the best course of action in a given terrain. The peer-production licence entails a specific kind of closure, one that aims to prioritise worker-owned approaches over shareholder-based capitalism. In fact, all forms of openness imply enclosures: from copyleft clauses in open source licences that force re-users to licence their works under the same conditions, to the legal requirement to attribute the creator of a CC BY-licensed work, to social norms around the use of public domain materials[3]. These are all forms of antagonism.

But antagonism implies a hegemonic struggle composed of conflicting power relations between groups with different points of view. Hegemony itself presupposes what Laclau and Mouffe describe as "the incomplete and open character of the social" (13, 134). Democracy is framed as a process of constant reinvention, but with a pluralistic, open character. Whereas the neoliberal response to openness is to enshrine it

---

3   See the Open Knowledge Foundations 'Open Definition' for more: opendefinition.org/od/2.1/en/

within the instruments of market-based measurement and logic, Mouffe and Laclau on the contrary argue that conflict and plurality actually constitute the very possibility of democracy—"If there is politics in society it is because there is conflict" (4). Democracy therefore requires institutions that promote plurality and difference.

For Adema and Hall, the development of democracy as a process parallels the development of open access. They argue it is helpful to think of open access "less as a project and model to be implemented, and more as a process of continuous struggle and critical resistance[.]" (1). Openness (and open access specifically) therefore implies a plurality of approaches and values; it is temporary, constantly changing and cannot be decided in advance.

This is why one sees a diverse range of projects operating under the "open" banner, not just those adopting a political approach one way of the other. It is also why one sees a number of projects operating in direct opposition to neoliberal approaches within publishing, libraries and the academy more generally, such as Punctum Books, Open Humanities Press and the Radical Librarians Collective. With so many diverse approaches espousing the open philosophy, perhaps this means that "the open" has a more complicated relationship with the political than meets the eye.

In this section I have shown that openness does have a real basis outside of neoliberalism and that it is an approach or process that requires careful articulation in an undecidable terrain. Although there are many "open" projects that do conform neatly to the neoliberal values of measurement by the market, there are many that do not and many that oppose it. The numerous motivations for various open access projects will become clearer in the second section of the chapter. Suffice it to say that this section does illustrate a lineage between open source and open access, particularly as many open access projects that evolve from open source culture focus on the potential of reuse, collaboration and remixing. However, not all understandings of "open access" derive from this lineage and instead reflect more of a preoccupation with the provision of public access to the research literature. What the genealogy does reveal is that openness requires one to consider the enclosures made in releasing something to a particular community, and this involves accepting the incomplete and contingent nature of things.

The next section aims to highlight the lineage of open access that stems primarily from the promotion of free access to research, as opposed to being primarily concerned with openness and reuse. Here it will be clear that the different approaches and motivations for open access reveal its individual, community-specific nature, leading to the conceptualisation of open access as a boundary object. This will ultimately illustrate that it is not possible to talk about open access as one thing, or even a thing-in-itself at all, but rather as a series of experiments in critical engagement with publishing processes, free culture and scholarly communications in general.

Access to research

Aside from OA's lineage from open source culture, and "openness" more broadly, there is also the parallel development of open access as derived from the desire to provide access to research to those who do not have it. This access-focused lineage does not necessarily require any separate approach to copyright or relaxed reuse permissions, the kinds of which are embedded in understandings of open access that derive more from open source software and free culture. The kinds of open access that prioritise access are often more conservative in their approach to research articles/books as fixed objects with traditional notions of authorship, rather than open notions of adaptation and remixing. The emphasis here is on simply removing price restrictions to a research work.

Forms of open access that prioritise access are often, though not exclusively, associated with repositories. An early example of this kind was the arXiv, which formalised the pre-existing culture in physics of sharing working papers (preprints) as soon as they were ready, prior to peer review and publication in a journal. High-energy physics always had a culture of sharing working papers—this was originally conducted via post and then by email to an exclusive list of "A-list" researchers (7, 3). The arXiv ensured that anyone with access to the internet could read cutting-edge physics research.

However, as arXiv founder Paul Ginsparg notes, the internet was "something of a private playground for academics, subject to few intrusions from the outside world" and so editorial and access controls were not necessary (7). This implies that if the "outside world" were more present on the early Web then the arXiv might not have been freely accessible to all. It seems likely then, despite its importance and success as a repository of publicly accessible physics and mathematics research, that its facilitation of access to knowledge was a by-product of the arXiv's original intentions. Rather, the arXiv increased the speed of dissemination of high-energy physics research to those whose access was delayed because they were not on the "A-list".

The arXiv is an example of an approach to open access that made pre-existing research dissemination practices more efficient. It worked within the constraints and affordances of what Karen Knorr Cetina terms high-energy physics' epistemic culture—"those amalgams of arrangements and mechanisms—bonded through affinity, necessity and historical coincidences—which, in a given field, make up *how we know what we know*"(12, 1). It is unlikely that the "open access" status of the arXiv was relevant or even noticeable to early users of the repository, if only because there were initially so few users on the Web. Its success was largely down to how it improved the existing research practices of high-energy physics researchers.

Implicit in both the arXiv and institutional repositories is the idea that research objects can be shared more effectively via digital technologies. This is the kind of access emphasised by John Willinsky's "access principle": A commitment to the value and quality of research carries with it a responsibility to extend the circulation of such work as far as possible and ideally to all who are interested in it and all who might profit by it (19, xii). The access principle describes a researcher's "responsibility" to disseminate their research to all who wish to read it. This is a foundational argument for open access: digital technologies enable a more effective way of sharing research such that everyone with a stable internet connection should be able to access it. It is an argument based on technology as an enabler of new or more efficient practices.

Arguments of this kind are often framed as a response to prohibitively high journal subscriptions, especially the "serials crisis" that affects academic libraries, referring to the increase in the price of journals above inflation such that increasingly few libraries can afford all the resources they need (See e.g., 17). In this instance, open access is a response to publisher pricing strategies and the perpetuation of a business model based on print rather than digital economics. Open access should therefore ease library budgets and have a positive effect both inside and outside the university.

Though the motivations here are numerous, arguments of this kind do not rely on the need for research to be libre (i.e., open source and reusable in commercial contexts). Repositories do not generally carry the requirement for articles to be uploaded under particular Creative Commons licenses. They are therefore associated more with gratis (free) access simply because this is sufficient to solve the original problem framed as a lack of access to research outputs. But there is a tension here between gratis access to research and what many believe is the canonical definition of open access: the Budapest Open Access Initiative definition, a minimum criteria of which is that open access should necessarily entail the ability to reuse a research paper. The only restriction should be to "give authors control over the integrity of their work and the right to be properly acknowledged and cited"[4]. Repositories in general do not provide the kind of open access that conforms to this definition.

This section has only scratched the surface of the various ways in which open access resonates differently within different communities. There are innumerable understandings of and motivations for providing access to research, from early web-based journals to contemporary book-based publishing houses, but their chief motivation is for getting more eyes on research. In this regard, open access for many means free-to-access research as opposed to any commitment to "openness" or reuse. This has important implications for the genealogy of open access and helps explain a number of features with the current ecosystem.

Open/Access as a boundary object

Having shown that open access is itself an approach or process with no fixed meaning or definition, it is helpful to theorise open access as a boundary object, a term first defined by Star and Griesemer in 1989. Boundary objects are physical or conceptual objects that are understood differently within individual communities but maintain enough structure so as to be understood between communities. As the authors write:

"Boundary objects are objects which are both plastic enough to adapt to local needs and the constraints of the several parties employing them, yet robust enough to maintain a common identity across sites. They are weakly structured in common use, and become strongly structured in individual site use. These objects may be abstract or concrete. They have different meanings in different social worlds but their structure is common enough to more than one world to make them recognizable, a means of translation" (15, 393).

It is the plasticity of the boundary object that is key. Boundary objects maintain a recognisable structure across communities despite being understood differently in different situations and contexts. Their structure is always open to change.

---

4   Budapest Open Access Initiative | Read the Budapest Open Access Initiative. http://www.budapestopenaccessinitiative.org/read

In terms of open access, if we accept that open access itself has a number of individual motivations and understandings, then it is best conceptualised as a boundary object. This means that open access resonates differently within individual communities of practice, not just within disciplinary communities but cross-disciplinary interest groups or those sharing a common methodology (or any community of practice, for that matter). It also allows open access advocates to share a common language despite not having a common vision or explicit shared understanding of what they are advocating.

However, as is well known, arguments over the correct definition of open access and strategies for how it should be pursued are rife within open access. Boundary objects, as Isto Huvila explains, do not escape the kinds of hegemonic struggles between perspectives. Boundary objects are not purely consensual and still rely on the need to make decisions or enclosures as to what the object represents. As Huvila argues: "the creation or reshaping of boundary objects is always an attempt to make an hegemonic intervention" (10). These kinds of hegemonic interventions are common throughout open access, especially around routes to open access, how open access should be funded, what licenses are required and whether top-down policies are needed.

It is clear that the conditions that enable OA's existence arise out of the two lineages between open source/free culture and access to research. Martin Eve makes a similar point, arguing that open access emerges at the "convergence point of these two narratives—problems of supply-/demand-side economics and the birth of the free culture movement". This is certainly a good way of framing the conditions for the possibility of open access, although one would not want to emphasise too much of a consensus between the two lineages. John Willinsky, for example, goes as far as to say that there is a "common cause" that unites open source, open access and open science, that the convergence of circumstances is in fact a convergence of intentions (18). Whereas my analysis illustrates that this is not always the case.

To speak of a "common cause" is to assume a fixed solution to a specific problem, but we have already seen that open access is neither of these things. Theorising it as a boundary object allows us to conceptualise open access as a community-led process without fixed meaning and continually open to interpretation. This will allow a number of individual experiments in openness to blossom, thus working against enclosure by any particular group. The important thing is that the diversity of approaches makes open access useful, rather than is enclosure at a general policy level.

Open access should therefore remain complicated and embrace the "undecidability" that Mouffe and Laclau reference. This would entail an ecosystem of experimental, community-governed projects based on articulated and unique approaches to openness, free culture and the gifting of one's research. Such messiness, if promoted as valuable in itself, would provide a space for diversity and the more marginalised voices and elements of academic research to be heard. It would also allow open access to not be so easily captured by dominant and/or neoliberal approaches based on high APCs paid to commercial publishers simply for the sake of satisfying ill-conceived government and funder mandates.

The genealogy: what have we learned?

This chapter has shown that open access has a complex genealogy and cannot be portrayed as a coherent or homogenous "movement". Not only are there two separate lineages of open access originating from "openness" on the one hand and access to research on the other. Within these lineages there are numerous motivations and understandings of the term. From Nathaniel Tkacz's analysis of openness we have seen that forms of open access may indeed reflect a neoliberal philosophy in the same way that forms of open source software sometimes do as well. Yet, I have also shown that open access is not always a neoliberal project and encompasses a variety of political, social and disciplinary motivations that cannot be reduced to one particular understanding.

It is worth stating I do not intend to make a strong value judgement about different routes to and motivations for open access, only to say that the genealogy of open access shows it to be conceptually multiple— best conceived as a boundary object. This is a crucial part of its value: open access represents a multitude of positions and strategies but is generally recognisable across cooperative communities of practice. This level of diversity should be instantiated in any ecosystem of open access through experimentation and tolerance of disagreement.

What this means is that open access cannot be painted with a broad brush as a single movement or project that a given group of advocates are trying to implement. Some voices shout louder than others, and others are better at influencing policy, but this should not be confused with a homogenous community of zealous advocates all pulling in the same direction, as many argue (e.g., 8 and 3). Similarly, open access is not best conceived, as Daniel Allington characterises the advocate position, as a "single purported solution" to one or many problems (2). Open access represents a number of approaches and motivations, some thought through better than others, and it is easy for critics to portray a particular approach to open access as representative of all of it.

For it to be politically progressive, the conditions for the adoption of open access should reflect and be answerable to the various communities of practice that conduct and publish research. The important thing here is for funders, institutions and governments to back away from implementing restrictive mandates and instead facilitate experimentation governed by communities themselves. A lot can be learned from ideas around collective governance of the commons and further experimentation and research is required in this area. We might look to any number of scholar-led initiatives to understand the choices made around such initiatives and how projects can be mutually supported and co-reliant. Of course, this would entail the need for greater power for communities over publishing infrastructures for books and journals, working with university presses and libraries to improve researcher governance of publishing. It would also require researchers to wrest back control of technical infrastructures such as repositories and academic social networks from commercial providers. This is no doubt not easy, but nonetheless certainly worth striving for.

References

(1) Adema, Janneke, and Gary Hall. 2013. 'The Political Nature of the Book: On Artists' Books and Radical Open Access'. New Formations 78 (78): 138–56.

(2) Allington, Daniel. 2013. 'On Open Access, and Why It's Not the Answer'. http://www.danielallington.net/2013/10/open-access-why-not-answer/.

(3) Beall, Jeffrey. 2013. 'The Open-Access Movement Is Not Really about Open Access'. TripleC: Communication, Capitalism & Critique. Open Access Journal for a Global Sustainable Information Society 11 (2): 589–597.

(4) Carpentier, Nico, Bart Cammaerts, and Chantal Mouffe. 2006. 'HEGEMONY, DEMOCRACY, AGONISM AND JOURNALISM: An Interview with Chantal Mouffe'. Journalism Studies 7 (6): 964–75.

(5) Foucault, Michel. 1984. 'Nietzsche, Genealogy, History'. In The Foucault Reader, edited by Paul Rabinow, 1st ed, 76–100. New York: Pantheon Books.

(6) Foucault, Michel. 1995. Discipline and Punish: The Birth of the Person. Translated by Alan Sheridan. New York: Vintage Books.

(7) Ginsparg, Paul. 2011. 'It Was Twenty Years Ago Today.' CoRR abs/1108.2700. http://arxiv.org/abs/1108.2700.

(8) Golumbia, David. 2016. 'Marxism and Open Access in the Humanities: Turning Academic Labor against Itself'. Workplace: A Journal for Academic Labor, no. 28.

(9) Hall, Gary. 2008. Digitize This Book!: The Politics of New Media, or Why We Need Open Access Now. Minneapolis: University of Minnesota Press.

(10) Huvila, Isto. 2011. 'The Politics of Boundary Objects: Hegemonic Interventions and the Making of a Document'. Journal of the American Society for Information Science and Technology 62 (12): 2528–39.

(11) Kleiner, Dmytri. 2010. The Telekommunist Manifesto. Amsterdam: Institute of Network Cultures.

(12) Knorr-Cetina, Karen. 1999. Epistemic Cultures: How the Sciences Make Knowledge. Cambridge Mass.: Harvard Univ. Press.

(13) Laclau, Ernesto, and Chantal Mouffe. 2001. Hegemony and Socialist Strategy : Towards a Radical Democratic Politics. London: Verso.

(14) Mouffe, Chantal. 2013. Agonistics: Thinking the World Politically. London: Verso.

(15) Star, Susan L., and J. R. Griesemer. 1989. 'Institutional Ecology, 'Translations' and Boundary Objects: Amateurs and Professionals in Berkeley's Museum of Vertebrate Zoology, 1907–39'. Social Studies of Science 19 (3): 387–420.

(16) Tkacz, Nathaniel. 2014. Wikipedia and the politics of openness. Chicago: University of Chicago Press.

(17) University of Illinois Library at Urbana-Champaign. 2009. 'The Cost of Journals'. University of Illinois Library at Urbana-Champaign. http://www.library.illinois.edu/scholcomm/journalcosts.html.

(18) Willinsky, John. 2005. 'The Unacknowledged Convergence of Open Source, Open Access, and Open Science'. First Monday 10 (8).

(19) Willinsky, John. 2006. The Access Principle: The Case for Open Access to Research and Scholarship. Cambridge, Mass.: MIT Press.

The Paradox of Success

Joachim Schöpfel

In 25 years, open access, i.e. free and unrestricted access to scientific information, has become a significant part of scientific communication. However, its success story should not conceal a fundamental change of its nature. Open access started, together with the Web, at the grassroots as a bottom-up, community-driven model of open journals and repositories. Today the key driving forces are no longer community-driven needs and objectives but commercial, institutional and political interests. This development serves the needs of the scientific community in so far as more and more content becomes available through open journals and repositories. Yet, the decline of open access as a community-driven model is running the risk of becoming dysfunctional for scientists and may create new barriers and digital divides.

A kind of Open Access blues

Never before has access to research been as open as in 2017. The most recent figures on the global growth of open access, whether through archives or publications, are without historical parallel: more than 112m documents are available through the Bielefeld Academic Search Engine, about 60% open access from more than 5,700 providers; the OpenDOAR database contains more than 3,300 repositories; and the Directory of Open Access Journals, even after removing 3,000 titles in 2016 and with a new and more selective policy, still counts 9,500 journals with 2.5m articles[1]. Also in 2016, the Member States of the European Union committed themselves to 100% open access to public research in 2020. More and more institutions and organisations apply mandatory policies. "Most publishers now offer open access options and publish open access journals, and work closely with funders, institutions and governments to facilitate these developments"[2]. Several studies confirm the advantage of open access to traditional publishing for scientists and their institutions in terms of impact, visibility and outreach (12).

And still, there is a kind of open access blues. Acceptance and uptake are slower than expected. Some historical figures of the open access movement have started to complain about a deficit of enthusiasm, a lack of grassroots support that diminishes the potential of network effects,[3] and about scientists who are "too lazy, too dim-witted or too timid" to provide green open access[4]. Are scientists "fools", "pedants" or "poltroons"[5], lacking awareness and knowledge? Are they too conservative or are they simply avoiding additional workload? To accelerate the movement, some open access proponents, such as Stevan Harnad, justify coercive strategies; institutions should force scientists for their own good, in particular through institutional green mandates. Can open access legitimately force people to buy into it for their own happiness? Simply put, does the end justify the means? Something has gone wrong. The 2017 version of open access has become different from what it should or could have been, namely the "beautiful vision of the potential unprecedented public good"[6].

Sometimes, the result of change differs from the initial goal. Sometimes too, the change is obtained for reasons other than the initial intentions. Open access started as a community-driven project but then changed into something entirely different. The turning point of the open access life cycle came in the early years of the 21st century when commercial publishers started considering open access no longer as a threat but as

---

1   For more insight and details, see the excellent site of Heather Morrison, University of Ottawa http://poeticeconomics.blogspot.fr/

2   STM publishers http://www.stm-assoc.org/public-affairs/resources/ publishers-support-sustainable-open-access/

3   Eric Van de Velde on http://scitechsociety.blogspot.fr/, July 24, 2016

4   Stevan Harnad in Global Open Access List goal@eprints.org, January 6, 2017

5   See http://openaccess.eprints.org/index.php?/archives/1198-FOA-Free-Online-Access.html

6   Heather Morrison loc.cit.

an opportunity. Institutional interests, commercial benefits and neoliberal ideology became key factors in the actual take-up of open access. These key factors may be useful for the success of open access, at least from certain vantage points. They do, however, carry the risk of new barriers to free and direct scientific communication.

Bottom-up

Open access started at the grassroots level. Scientists began to adopt internet technology for free and rapid dissemination of content in the late 80s. They did this by applying the new technology to the oldest and most functional type of scientific communication, i.e. the academic journal (5). *New Horizons of Adult Education* was launched in 1987, *Psycoloquy* in 1989, the *Electronic Journal of Communication*, *Postmodern Culture* and the *Bryn Mawr Classical Review* in 1990. Years before the term was coined, these titles invented the gold road to open access. They were academic and scholarly journals in digital format, peer reviewed or refereed, publishing original content, reviews etc. in a particular scientific field, and they were freely available on the internet, without charge to the reader.

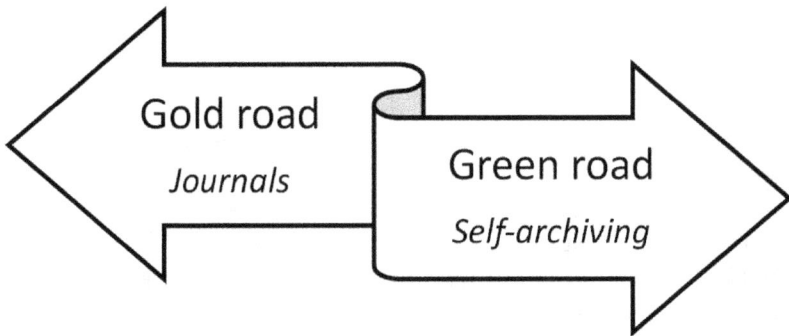

Figure 1. Gold and green road to open access

After gold came green (1). 1990 was a pivotal year for open access. Tim Berners-Lee published his *Proposal for a HyperText Project* called the WorldWideWeb, and the High Energy Particle Physics (HEP) community at the Stanford Linear Accelerator Center started to test direct communication of research papers on the internet, based on the format TEX (7).

However, the email exchange of attachments in TEX was rapidly jamming mailboxes. The junction between the CERN technology and HEP was made in 1991 with the launch of a Web service called arXiv on a Los Alamos server. Paul Ginsparg and his colleagues had invented the green road, i.e. self-archiving of scientific articles, conference papers etc. deposited in some kind of open (institutional, disciplinary...) repository.

The success of both models is well documented by the enlistment figures in international directories like OpenDOAR, ROAR and DOAJ (2)(11). Not all of the pioneering gold titles survived; not all of the early launched repositories are still operational and online. But in the following years, more and more open access journals were released, supported by initiatives like the Scholarly Publishing and Academic Resources Coalition (SPARC) and others, and a growing number of institutions and organizations introduced their own open repositories.

The figures referenced above are impressive, and the annual growth even more. Here are some examples provided by Heather Morrison, with figures representing the annual growth for the period 2016-2017 (with absolute numbers, posted on 30th June 2017):

- Internet Archive (texts): 27% (12.8m)
- BASE search engine (documents): 20% (112.5m)
- DOAJ directory (articles): 14% (2.5m)
- arXiv (preprints): 10% (1.3m)

ArXiv is still present. ArXiv is not only the first and best-known open repository but it is also the best example of the special characteristics of these precursor initiatives: it was not only the right technology in the right place and at the right time, but there was also (and above all) the fact that the HEP community at that time already had a long tradition of sharing resources not on an individual level but as a community. The key factor was (and still is) compatibility between technology and the community's already existing information and communication practice. A bottom-up project, one might say.

Despite the success of ArXiv bottom-up is not a guarantee of success. Several bottom-up journals were launched with much enthusiasm, personal commitment, and sometimes even with a degree of self-denial (scientists are not paid for publishing journals but for researching), and then discontinued, like *Psycoloquy*. Other types of bottom-up open access scientific communication appeared at that early "pioneer times" of the Web, in particular personal websites followed by blogs[7] and other social media. Their development produced a fundamental debate on the quality and functions of scientific communication. Scientists adopt communication vectors in so far as they serve their interests and needs. How can quick and large dissemination, including goals like impact optimization and social responsibility be reconciled with quality control, recording and preservation? These quickly became pressing questions.

In the late 90s, this debate was joined by the serials crisis, i.e. the widening gap between subscription costs and library budgets. The idea was simple and rooted in common law and historical experience: a concerted action by librarians and scientists voluntarily abstaining from subscribing and submitting papers to expensive commercial journals, as a form of protest and financial pressure. This was a grassroots action, equivalent to consumer activism. And it was more than that, since it included the proposal of an alternative non-profit, open communication system. This was not only a boycott but also a harbinger of system change.

The 2000 Public Library of Science (PLOS) online petition initiative by Michael Eisen, Patrick O. Brown, Harold E. Varmus and others was a big success. Tens of thousands of scientists and librarians all over the world signed it. This was the culminating point of the community-driven model of open access. But shortly afterwards, it became obvious that this grassroots action would be ineffective. Institutions (and libraries) continue to subscribe to commercial journals, and especially to the most expensive while because scientists—some of whom, presumably, had signed the PLOS petition—continued to publish in them.

Top-down

The final failure of the PLOS petition was not the end of open access but the beginning of the next chapter. And paradoxically, this new chapter appears to provide the key to the current success of open access. It took shape between 2000 and 2003. These were crucial years for open access, with new initiatives and projects, decisions, meetings and declarations. Three developments during this period contributed to the fall of open access as a community-driven model.

---

7   See for instance the French OpenEdition platform with more than 2,000 research blogs (carnets de recherche) https://www.openedition.org/catalogue-notebooks

Economic innovation: For some years, publishers considered open access to be a threat. But with new open access journals like BioMedCentral and PLOS One, shortly followed by the hybrid Springer journals, they discovered that open access offers new and even better opportunities than before in so far as the new business models with deals based on article processing charges paid by authors or their institutions (APCs) are no longer limited to constricted library budgets but could draw upon larger institutional research budgets. Today, open access "constitutes one of the few segments in the global publishing market that shows healthy growth rates" (3, p.13). "Models change, profits don't"[8]. Perhaps in the future, this process will be considered as a modern example of what Schumpeter termed "creative destruction".

Neoliberalism: Some of the main stimulants of the open access movement were actively initiated and supported by American charities (4). Their objective had one name: the "open society", in the tradition of Karl Popper[9]. Their enemies: barriers to free competition, such as socialist regimes in Central Europe and monopolies in capitalistic societies. Perhaps, and years ahead of their time, they considered the open access movement to be a kind of "orange revolution" in scientific publishing. These groups emphasized the green gratis open access with immediate-deposit mandates as this faces less legal restrictions, and thus could be more efficient in destroying monopolies. One thing must be said, however—private foundations, as charitable or as philanthropic as they may be, are not emanations of the scientific research community.

Evaluation: Funding bodies, research organizations and universities came to understand the benefits they can get from open access. Free and unrestricted dissemination of scientific information increases the institutional visibility and impact on the internet leading to improved international ranking and scientific competition. Also, institutional repositories can be much better tools for academic evaluation and control than library catalogues or bibliographic databases, provided their metadata are more or less exhaustive and linked to current research information systems (CRIS) or functionalities. Moreover,

---

8   Leti Kleyn on https://theconversation.com/, October 26, 2015

9   See Popper, K. R. (1945). The open society and its enemies. Routledge, London.

today, these CRIS have started to substitute institutional repositories, through the integration of their main functions. The institutional need and interest for independent monitoring and output assessment is a strong driver for green mandatory politics, often supported by academic librarians, notably less by the scientists themselves.

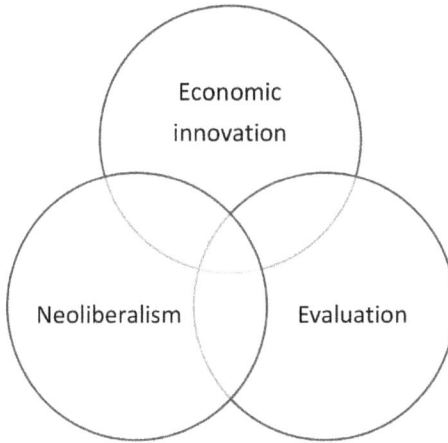

Figure 2. Three key factors of success

Economic innovation, neoliberalism and evaluation have in common the fact that they are not rooted in the scientific community, and that their main goals are not to satisfy scientists' needs for efficient communication but instead to maximize other interests. Nonetheless these forces can satisfy one part of scientist's needs, namely the goal of offering more freely available content. Other factors also contributed to the fall of the community-driven model of open access: the economic inefficiency of grassroots gold publishing and the need for consolidation and concentration to achieve economies of scale; the so-called citation-advantage of open access articles which not only serves as an incentive for authors to publish in open access journals but moreover, improves the impact and quality of gold journals and provides a formidable argument for marketing and sales; the public policy of open data pushing public research organizations "to go open"; and the need for significant investment in powerful and innovative services for the retrieval, selection, preservation and analysis of the overwhelming volume of open content. All of these factors conspired to favour the top-down approach.

Both sides now?

Open access to scientific information started at the grassroots, with projects and initiatives to adopt the new internet and Web technologies to satisfy some of the scientists' interests and needs, such as rapid and direct communication of selected (reviewed) content, widespread access, preservation and retrieval. 25 years later, the key drivers of the open access movement are no longer the scientists themselves but rather research managers, publishers, information professionals and politicians. This may explain some alienation between open access propoenents and research communities, which partly seem to consider open access as a new obligation and not a new opportunity.

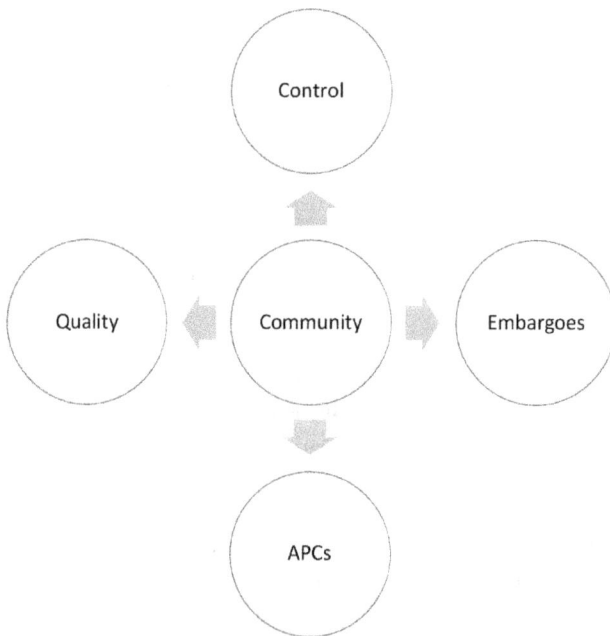

Figure 3. Potential conflicting interests

So far, the driving forces of commercial, institutional and political interests have contributed to the success of open access. The problem is that the nexus of these interests harbors the potential for conflicting interests that may be dysfunctional for the further development of open access, such as:

Control: Academic freedom is a "troublesome concept" (6). It may be hard to define but just as with any kind of freedom, one knows very well when it is at risk or is lost. Academic freedom (which research to do, how to do research, how to disseminate the results and so on) is an old concept, deeply rooted in the history of European universities, and the debate on science and institutional interests or "superior interests of society" is not really new. Scientists are anarchists, and states cannot be trusted. To gain meaningful support from scientists, open access advocates should not be opposed to academic freedom.

Embargoes: The main impetus or the initial models of open access was the immediate dissemination of scientific information. Embargoes protect commercial interests and may serve institutional interests, as they make near-to-exhaustive institutional repositories possible. But as this exhaustiveness is reduced to metadata, embargoes (and on-campus access) are in conflict with the scientists' need for direct and immediate communication. This conflict may reduce the benefit of institutional repositories for scientists. It may also reduce the relevance and impact of new laws on secondary exploitation rights with embargoes of six and twelve months, such as in France or Germany.

Article processing charges: Without doubt, APCs contribute to the success of gold open access. They serve commercial interests, they are compliant with institutional interests, and they can satisfy the scientists' needs for immediate and large dissemination of reviewed content (9), including the citation advantage mentioned above. The risk of this new business model is that it may create a new digital divide between rich research organizations and the others ("gold for the rich, green for the rest") and that it may even produce a second serials crisis because of a new gap between APC expenditures and cumulated budgets for scientific information. More transparency, i.e. denunciation of confidentiality clauses and disclosure of APC and offsetting conditions is the first step in controlling this potential conflict.

Dubious quality and information overload: Large open repositories and predatory publishing (a kind of collateral damage of gold open access based on APCs) are at the risk of producing open content of dubious quality, because of different versions, non-selected items, lack of peer review etc. Dubious quality is also a risk of repository metadata. Ever-increasing volumes of open content without adequate discovery tools are not helpful for the information needs of the scientific community.

Potential conflicts and dysfunctional developments are not necessarily bad as they may trigger new solutions. That being said, the current dynamics of scientific communication are contradictory and paradoxical. While the European Union adopted a Call for Action in favour of open science, a clear and consistent strategy is missing. Some countries—above all those with rich research organisations and universities—appear tempted by the APC-based gold road to open access while others rather support the green road, i.e. open repositories, and public investment in open access infrastructures. Apparently, the ever increasing number of freely available publications does not really reduce information inequalities and the digital divide between rich and poor research; large parts of research results remain behind the subscriber pay wall, many institutions cannot (or will not) afford APCs, and institutional repositories (still) do not satisfy the need for rapid and direct access to research.

If the situation was different (better), how then could the large success of the massive fraud on the Sci-Hub platform be explained? Besides, should Sci-Hub be considered as a community-based initiative to improve access to scientific information or as a copyright infringement that must be condemned? Or both? "Sci-Hub is obviously illegal," says structural biologist Stephen Curry at Imperial College London in the United Kingdom. "But the fact that it is so immensely popular, inside and outside academia, is a symptom of many people's frustration with the status quo in academic publishing." (8).

Like two decades ago, there is a growing debate among librarians and scientists on the boycott of commercial publishers and the cancelling of journal subscriptions (big deals) and—this is new—of how to avoid double dipping, i.e. paying APCs and subscriptions. Are we facing a new kind of serials crisis? Will these initiatives be more successful than before? Or are these just transitional problems on the road to open access and open science? The German Max-Planck-Society launched the OA2020 initiative in order to reboot the approach to open access by using subscription budgets to pay publishing fees (10). If this initiative were successful, it would change the business model but would at the same time cement the dysfunctional STI market and guarantee for more time the unacceptably high profits of certain large academic publishers. Is that the price to pay for open access? Is achieving open access to the formal published record worth more than all other goals? In France, a group of scientists and information managers responded to the OA2020 initiative with an appeal in favour of innovative forms of academic publishing—specifically, a call for public investment in new infrastructures and platforms for the open dissemination of research results beyond the STI market ("Appel de Jussieu").

In the early years of the movement, proponents considered open access to be a revolution of scientific publishing, with the potential to change the system. So far, open access has actually consolidated the pre-exisitng system, fostering further concentration along with the emergence of some new players. Will open access someday devour its own tail? Will it become a new administrative burden that supports a more efficient commercial system of scientific communication, by responding to political and institutional standards of direct and free access to research results? Or will institutions and people, scientists and professionals be able to realise the "beautiful vision" mentioned above without disregarding the particular needs and practices of scientists themselves? The questions lay before us, and the answers are open.

References

(1) Björk, B.-C., et al. (2014). Anatomy of green open access. *Journal of the American Society for Information Science and Technology* 65 (2), 237–50.

(2) Das, A.-K. (2015). Open access: History and developments. In *Introduction to Open Access*, p. 17–30. Paris: UNESCO.

(3) Fund, S. (2017). Take-off for OA books. *Research Information* (88), 12–13.

(4) Guilhot, N. (2007). Reforming the world: George Soros, global capitalism and the philanthropic management of the social sciences. *Critical Sociology* 33 (3), 447–77.

(5) Laakso, M., et al. (2011). The development of open access journal publishing from 1993 to 2009. *PLoS One* 6(6), e20961+.

(6) Neylon, C. (2015). Freedoms and responsibilities: Goffman, Hunt, Bohannan and Stapel. *Science in the Open*, 24 June 2015.

(7) O'Connell, H. B. (2000). Physicists thriving with paperless publishing. In *AAAS Meeting, February 2000. Washington, DC*.

(8) Schiermeier, Q. (2017). US court grants Elsevier millions in damages from Sci-Hub. *Nature News*, 22 June 2017.

(9) Schimmer, R., K.K. Geschuhn, and A. Vogler, (2015). *Disrupting the subscription journals' business model for the necessary large-scale transformation to open access. A White Paper*. Max Planck Digital Library, Munich.

(10) Schimmer, R. (2017). The transformation of scientific journal publishing: Open access after the Berlin 12 conference. *Information Services & Use 37* (1), 7–11.

(11) Suber, P. (2015). Timeline of the open access movement. *Open Access Directory*.

(12) Wang, X., et al. (2015). The open access advantage considering citation, article usage and social media attention. *Scientometrics 103* (2), 555–64.

This chapter is a revised and up-dated version of Schöpfel, J. (2015). Open access—the rise and fall of a community-driven model of scientific communication. *Learned Publishing* 28(4), 321–25.

Open Access and Symbolic Gift Giving

Ulrich Herb

Open access has changed. At the beginning of the millennium, it was portrayed in a romanticizing way and was embedded in a conceptual ensemble of participation, democratization, digital commons and equality. Nowadays, open access seems to be exclusive: to the extent that commercial players have discovered it as a business model and article fees have become a defining feature of gold open access, open access has increasingly transformed into a distinguishing feature and an exclusive element. Scientists are beginning to make the choice of a university or research institution as an employer based on whether or not these can afford to cover the article fees for publications in high-impact but high-priced journals. Surprisingly, this transformation of open access is not the subject of any noteworthy discussion in specialist or journalistic publications, but instead the ideals of the digital commons of knowledge still prevail in these venues. Even so open access is increasingly becoming an instrument that creates exclusivity, exclusion, distinction and prestige. These functions, however, are obscured by symbolic gift giving strategies and presented as altruistically staged, so that in the discourse of the open access community and in media reporting on open access, the both euphemistic and largely obsolete prosocial story-telling of open access dominates. The paper also discusses the question of whether the concept of open access was not overstrained by the hopes placed in it.

Open Access 2002: revolution, romance & idealism

In its early days, open access was mainly driven by altruism. The concept of making scientific knowledge available to everyone at no cost arose out of and was embedded in a morally motivated framework encompassing the idea of a digital knowledge commons, revolution, the levelling of knowledge-based differences, and democratization.

Even 15 years later, this moral impetus can be felt when we read the central passages of the Budapest Open Access Initiative's (BOAI) declaration: "Removing access barriers to (…) [scientific] literature will accelerate research, enrich education, share the learning of the rich with the poor and the poor with the rich, make this literature as useful as it can be, and lay the foundation for uniting humanity in a common intellectual conversation and quest for knowledge" (10). Open access was expected to make the world fairer, it was praised as a "democratizing tool that equalizes standards and expectations between lesser and greater institutions of learning, regardless of social rank or geographic location" (2, p. 4). Especially at the beginning of the millennium open access advocates were convinced to live in the era of a radical change: Harold Varmus, co-founder of the open access publisher PLOS (Public Library of Science) declared open access a "Revolution in the Publication of Scientific Papers" (28). Whoever published a paper in PLOS Biology was even lifted into the exalted position of a leader of a revolution: "We hope that you will lead the Open Access revolution by publishing your most exciting research in PLOS Biology" (5). David Prosser, then Director of the Scholarly Publishing and Academic Resources Coalition (SPARC), saw open access as "no less than the next information revolution" (27).

Revolution?

In 2017, this early enthusiasm has given way to a dry pragmatism; open access is primarily defined as gold open access and is, in large part, driven by the well-known players from the subscription business who are now publishing numerous journals which promulgate articles in exchange for Article Publication Charges (APCs). On August 1, 2017, the Directory of Open Access Journals (DOAJ) listed 9,621 journals. Only 32 publishers put out more than 20 journals, which gave them a quantitatively significant influence on open access.[1] These 32 publishing outlets published a total of 2,950 journals, or 31 % of all DOAJ journals. On this day the DOAJ listed 7,474 publishers. This means that 0.43 % of DOAJ publishers bring out 31 % of all journals. Of these 2,950 journals, 1,641 (or 56 %) originate

---

1   PLOS is not among those, but nonetheless exerts influence, albeit of a more qualitative nature.

from publishers that dominate the subscription market: Elsevier, Springer Nature (including BioMed Central, Frontiers) , Wiley, SAGE, De Gruyter, Taylor&Francis, Oxford University Press, Wolters Kluwer (15). Regarding the number of open access journals, Elsevier is the biggest open access provider (20) and findings from project OpenAPC[2] reveal that between 2005 and 2015, the bulk of APC-based articles in Germany were published by Springer Nature (17). These figures indicate that the much-touted revolution for the publishing market did in fact not happen, and maybe we can even conclude that it **could never have happened**, as we shall see later.

Democracy? Levelling Differences?

The early enthusiasm notwithstanding, soon after the inception of open access it appeared more and more questionable whether open access could indeed fulfill the expectations and hopes placed on it, e.g. the assumption that an open availability of scientific information would level social inequalities, optimize education or boost democracy was critizised as being very simplistic and not backed up by Sociology (14). Especially with regard to the cost-free availability of information, it could be doubted as early as 2010 if such free access to information could really achieve a levelling of social inequalities. The possibility of a fruitful application of available information depends first on the extent of a person's cultural capital, or more simply put, on that person's education (14). Open access does not change this reality. Constraints on the ability to use freely accessible information for one's gain and betterment are exerted by a person's cultural, economic and social capital.

The Indian city of Bangalore is a case in point: here, digitized land registry information was made available free of charge to everyone—without observing any ensuing levelling of differences or inequities. Instead, the cultural, economic and social capital as defined by Bourdieu (7) displayed its full force. The beneficiaries were found in the educated and wealthy segments of the population, given that they already had the requisite cultural and economic capital and only needed this newly available information about the real estate market to further increase their economic gain (4). Gurstein (13) sums up the findings: "The newly digitized and openly accessible data allowed the well-to-do to take the information provided

---

2   OpenAPC wants to offer a cost-monitoring for gold open access publications.

and use that as the basis for instructions to land surveyors and lawyers and others to challenge titles, exploit gaps in titles, take advantage of mistakes in documentation, identify opportunities and targets for bribery, among others. They were able to directly translate their enhanced access to information along with their already available access to capital and professional skills into unequal contests around land titles, court actions, and offers of purchase for self-benefit and to further marginalize those already marginalized." Also, the mere presence and accessibility of technical innovations, internet connectivity and information of all sorts will not—by itself—miraculously do away with social inequalities, uneven distribution of privileges and disenfranchisement (see e.g. 16).

Provisional Conclusion

In 2017 it is clear that open access could not fulfil the expectations and hopes of the revolution placed on it. Nor could the other positive structural effects predicted by open access advocates of 2002 such as a furthering of democracy and cultural and economic levelling be realized. Given that open access was not able to fulfil these promises, we also must ask if there were negative or dysfunctional effects brought on by open access. To answer this question we can again consult French sociologist Pierre Bourdieu and his theories of social fields (especially for the field of science see 8) and distinction (6).

Social Fields and Distinction

Bourdieu's theory of fields says that individuals and institutions are acting in *social fields*. Bourdieu describes fields as universes, encompassing actors and institutions, with more or less specific social rules (14). These fields are mostly vertically stratified and structured areas of competition. The actors within these fields, individuals and institutions, pursue strategies and agendas—mainly to achieve power and distinction, e.g. by accumulating social, cultural, economic or symbolic capital, in order to distinguish themselves positively from other actors in the field.

The different types of accumulated capital structure the different social fields and are used to achieve and obtain distinction and distinctiveness. The structures formed by the distribution of this capital dictate the rules for the various fields, and all three factors—distribution of capital, structures and rules—are marked by persistence. These mechanisms are active in all social fields, also in science and academia as well as in academic publishing.

But what do these theories teach us about the instrumentalizations of online publishing and open access? The internet is a condition sine qua non of open access—but it is nothing more than a technical[3] infrastructure—and open access is nothing more than an alternative way of publishing. Therefore it might be naïve to think that it could change these inherent characteristics of social action and fields. Even worse: It should be expected that open access and the internet are nothing more than new instruments or gadgets that will be utilized by the actors in the field of science in a very well-known way: They will be used "to gain and raise (…) reputations, build exclusive groups and exclude others. No matter how *open* a network or infrastructure [addition by the author: or principle] might be, they are devised, shaped and used by individuals. Consequently, their utilization is subject to human interests and necessities like networks constructed for distinctiveness and power" (14).

Exclusion

As stated before, distinction and distinctiveness are often reached by means of exclusion. But how could a project launched with the expressly stated goal of uniting humanity morph into a program deeply exclusionary in character? A glance at the history of open access helps to understand this unexpected turnaround, which is connected to the definition of political goals.

In its early years, open access was defined, shaped and driven forward primarily by scientists, librarians and research funders; both green and gold open access were seen as being on par with each other. Yet these actors envisioned different aims for open access:

- Research funders were interested in greater impact and dissemination of funded research
- Scientists were aiming for cost-free access to the publications of their peers as well as for greater impact and more dissemination for their own publications

---

3   One might add: a hierarchical infrastructure.

- Librarians were hoping to ease pressure on their budgets posed by the increasing costs for journal subscriptions.

Curiously, one group of actors, the traditional courier of the academic and scientific publication system, was absent from the discussion about open access: the commercial publishers[4]. Their absence had a reason: The BOAI does not mention anything about the profit potential of open access but focuses instead on disinterestedness and altruism[5]—and was thus of no interest to the publishers.[6] Furthermore, scientists, librarians and research sponsors also regarded open access as a panacea for the exploding costs, the manifestations of the profit motives of the academic and scientific publishers; both camps, libraries, researchers and research funders on the one hand and the publishers on the other remained in opposition to each other. However, not all scientists embraced open access readily and with open arms. Arguments against green open access included unsettled legal issues about copyright, the toil to deposit a file different from the publisher's version on a repository as well as the preference for the Version of Record rather than a download from the repository. The chief argument against gold open access was the perceived lack of reputable open access journals. The promise of academic reputation, standing among one's peers, and advancing career prospects appeared to be more successfully realized by the commercial publishers.

Despite this reluctance within the research community regarding open access, research funders and scientific organizations issued lofty quantitative goals for open access. The European Union's Competitiveness Council aims to make at least 60 % of publications available in open access by 2020, with the full 100 % of publications envisioned for 2025 (12).

---

4 The declaration of the BOAI mentions publishers only once; *commercial* publishers are even not mentioned explicitly.

5 References to money are stated only in negations („without payment", „remove the barriers, especially the price barriers", „without financial … barriers") as if there were no place for monetary interests in Open Access. Although there is mention of 'costs needing to be covered' there are no references to profits („open access is economically feasible", the overall costs of providing open access … are far lower than the costs of traditional forms of dissemination") (10).

6 Another obstacle: The commercial publishers were in the middle of the digital transformation and fought to bring their sales model into line with the new internet economy. Open access as a principle of free document use overstrained them in 2001; the shift was marked by the years 2006–2008. Springer, for instance, acquired the Open Access publisher BioMed Central in 2008.

Goals as ambitious as these cannot be reached via green open access alone but require the large-scale utilization of gold open access. This is reflected by the fact that research sponsors increasingly pay the (frequently unlimited) APCs for publications in open access journals. This avenue is further strengthened through the issuance of national licenses for subscription journals which allow open access publishing in such journals, as exemplified by the nation-wide licence between the Netherlands and both Springer Nature and Elsevier or between Austria and Springer Nature. Similar agreements of this nature are to be expected for the future.

The outcome of this strategy is three-fold: on the one hand, open access becomes an increasingly lucrative option for commercial science publishers; for reluctant researchers, on the other hand, it is an increasingly attractive publication outlet because they can now publish their work via open access—for a fee—in prestigious journals of their field.[7]

The third outcome is an increasing exclusivity observed especially at international open access conferences: attendance at the Berlin 12 Conference, for example, a yearly open access conference was an invitation-only affair, and a list of the participants was not made public even after repeated requests (26). The conference website issued only a brief explanation for this: "The 12th conference in the Berlin Open Access Series will be an invitation-only workshop for high-level representatives of the world's most eminent research organizations. (...) The central theme will be the transformation of subscription journals to Open Access" (8).[8] The Open Access Amsterdam Conference in 2016 was formally open to everyone, but the target audience addressed in reality by the conference organizers was very different from the enthusiasts of the early open access era: "The venue will be the spectacular building of the Royal Tropical Institute in Amsterdam. Hundreds of scientists, entrepreneurs, publishers and global thought leaders will come together to further the objectives of

---

7   This avenue is further strengthened through the issuance of national licenses for subscription journals which allow open access publishing in such journals, as exemplified by the nation-wide licence between the Netherlands and both Springer Nature and Elsevier or between Austria and Springer Nature. Similar agreements of this nature are to be expected for the future.

8   On the first day of the following Berlin 13 Conference participation was reserved for the signatories of the OA2020 "expression of the interests" (which consider the described transformation of closed access journals as the way to promote open access) and for observers from scientific institutions. The second day was open to all interested parties; especially representatives of the publishing industry were invited. (19)

Open Access and to discuss the importance of free knowledge sharing in the innovation processes of the interconnected world." (24). The exclusivity of the venue, the selection of the addressees (apart from researchers, publishers and 'thought-leaders' and also entrepreneurs—all representing the commercial forces within open access), and finally, the hefty conference fee of 475 € all work in tandem to thwart and foil the original participatory essence and thrust of open access. The privilege of discussing the importance of free knowledge with key players is thus conferred only on actors with deep pockets. In the same vein, the Berlin 13 conference focused solely "on the large-scale transformation of scholarly journals from subscription to open access" (23; this statement is underpinned by the conference's agenda, see 22).

Symbolic Goods and Symbolic Gift Giving, Excellence and Elitism

The envisioned transition toward widespread gold open access definitely sounds the death knell for the open access revolution, as the market share of the already dominant commercial publishers will be further consolidated. The preference for commercial open access will also introduce an aspect of privilege and excellence into the project: tying open access publication to existing licensing systems turns open access into a privilege since researchers from wealthy nations that can afford to pay for national licenses derive verifiable and tangible benefits from high citation figures in open access documents (see e.g. 3)—whereas less well-heeled countries not only have to pay high subscription fees for journals but also APCs for open access publishing. In practical terms, an open access option underwritten by subscription schemes amounts to a cost rebate for the institutions of nations willing and able to pay for these schemes.

Seen from a sociological perspective, investing in APCs is similar to symbolic gift giving in that an institution—through the payment of these fees—gives the community access to articles published by that institution itself. This kind of exchange is a frequent substitute for a "formal anonymous market" (11, p. 181) in areas where such a market does not exist, as, for instance, in the academic world. Invariably, symbolic gift giving is also a demonstration of one's own potency and "helps clarify social roles, wealth, or status" (11, p. 181). Making available a lot of (expensive) open access is therefore an effective way for an institution to highlight and underscore their exalted position in the realm of science and research.

In the terminology of Bourdieu (6, p. 66), the privilege of publishing in high-APC open access journals is a symbolic good, and "the manner of using symbolic goods, especially those regarded as the attributes of excellence, constitutes one of the key markers of 'class' and also the ideal weapon in strategies of distinction." Consequently, universities and research institutions capable of creating resource pools for covering the APC fees for open access publishing—sometimes to the tune of more than 9,000 € in APC fees per publication[9]—distinguish themselves favorably from their competitors and send a clear signal to young and aspiring researchers they want to attract. So open access becomes more or less a "luxury" and will "increase competition in an already highly competitive funding regime" (29, p. 58).

In the mid-term future, this competition can be expected to result in clear cumulative effects such that publications in highly-ranked—and high-APC—journals ensure the high citation rates that are attractive to the much sought-after academic stars. At the same time, the positive outcome for the university consists in higher impact scores and resource allocations to their budgets and more project approvals by funders. As this process becomes self-perpetuating, the competition between universities will increasingly be limited to fewer and fewer actors and eventually be closed completely for all but a small number of elite institutions—or as Weller puts it: "Ironically, openness may lead to elitism" (29, p. 58).

This conclusion is reflected in the findings of a study by Jahn and Tullney (17, p. 7 ff.) analyzing the APC fees paid by German universities and research institutions between 2005 and 2015: 39 % of APC-financed articles and 38 % of paid APC fees were generated by the Max-Planck-Society, the most prestigious German research institution. Looking at the numbers of APC-financed articles the following four places were also held by prominent research institutions: Göttingen University, the Karlsruhe Institute of Technology KIT, Regensburg University and Ludwig-Maximilians-University Munich.

Summing up

Of course it can be stated that cost-free or low-fee gold open access or green open access continues to exist. But the agenda of the leading research

---

9   As documented in the data provided by the project OpenAPC (1).

institutions, the relevant policy-makers, and the conferences receiving the highest media attention by now focus almost exclusively on the commercial version of open access, leading to an ever-greater preference for this type of open access. This unexpected (seen from the perspective of 2002) manifestation of open access as a business model and its instrumentalization for the production of exclusivity and distinction, sadly, is hardly ever discussed.

How is this shift away from the early mission of open access communicated to the world? In truth, hardly at all. Within the open access community, the development is mostly ignored or swept under the carpet; in the interest of maintaining unity, the old open access idealism is taken out and paraded about.[10] This type of romanticism is also used in the external communication to the media, especially when news and magazines' comments on Sci-Hub criticize commercial publishers dominating the subscription market (e.g. 21): Theses publishers are blamed for impeding the dissemination of scientific information by charging libraries with sky-rocketing subscription fees whereas gold open access is usually depicted the solution to this financial misery—whereas in actual fact, exactly these publishers are already actively shaping open access. But these romantic illustrations are no more than a rhetorical embellishment of open access policy and devoid of real substance.

Meanwhile, in science policy and media the striving for distinction and exclusivity through symbolic gift giving is staged as an act of selflessness. This selflessness is neither a deception nor is it truly selfless: "…at a subliminal level, the 'pure' and unselfish interest is an interest in selflessness, a kind of interest that is characteristic of the economy of all symbolic goods; in this economy, it is the unselfishness which carries the reward. Thus, in a certain way, the strategies of the actors are always two-faced, ambiguous, driven by interests as well as disinterested, inspired by a kind of unselfish self-interest which allows for completely antagonistic but equally erroneous (on account of their one-sidedness) description of motives— one hagiographic and idealizing, the other cynical and reductionist in its denunciation of a scientific capitalist as a capitalist like any other." (8, p. 24 f. translated by the author).

---

10   Exemplary the motto for events during the Open Access Week 2016 at Brunel University (London) may be quoted: "The Revolution Will Not Be Televised (but it may be tweeted)" (9).

Summing up we can say that open access did not disappoint all the expectations placed on it. Regarding research efficiency open access is a success: It speeds up scientific communication, makes science more transparent and verifiable, it facilitates the re-use of scientific information, and it generates higher impact scores. Nevertheless, it failed to realize the idealistic hopes connected with it. Maybe the path to that failure was predefined. After all, open access is only one variant of scientific publishing that could not possibly revolutionize the entire system. Besides, we should not forget: No matter whether we think about open or closed access, both are part of the field of scientific publishing that still consists of the same stakeholders with the same power (or without it), playing according to well-known rules and being subject to equally well-known interdependencies. (25).

Even the use of open access to create exclusivity is not really a surprise in hindsight. As Hilbert (16, p. 832) states with regard to other innovations, their use and application is always embedded in "an entire symbolic universe of social status". It is this embeddedness in social structures and rules that is responsible for the fact that the early idealism surrounding open access fell victim, in large measure, to the very success of open access in the realm of research efficiency. Open access 2018 is primarily defined by attributes such as outreach or impact that can be exploited to produce excellence and distinction—both of which can be bought for a fee from commercial publishers.

References

(1)   Ahlborn, B., et al. (2017). "Datasets on Fee-Based Open Access Publishing across German Institutions. Release 3.17.12." Bielefeld University. https://github.com/OpenAPC/openapc-de/tree/v3.17.12.

(2)   Arcadia Fund (2010). "Academic Knowledge, Open Access and Democracy." https://www.arcadia-fund.org.uk/academic-knowledge-open-access-and-democracy/.

(3)   Archambault, É., et al. (2016). "Research Impact of Paywalled versus Open Access Papers." http://www.1science.com/oanumbr.html.

(4)   Benjamin, S., R. Bhuvaneswari, and P. Rajan (2007). "Bhoomi : ' E-Governance ', or , an Anti-Politics Machine Necessary to Globalize Bangalore ?" A CASUM-M Working Papers. doi:10.1146/annurev.anthro.012809.104953.

(5)   Bernstein, P., et al. (2003). "PLoS Biology—We're Open." *PLoS Biology* 1 (1): e34. doi:10.1371/journal.pbio.0000034.

(6)   Bourdieu, P. (1984). *Distinction : A Social Critique of the Judgement of Taste*. Cambridge: Harvard University Press. https://monoskop.org/images/e/e0/Pierre_Bourdieu_Distinction_A_Social_Critique_of_the_Judgement_of_Taste_1984.pdf.

(7)   Bourdieu, P. (1986). "The Forms of Capital." In *Handbook of Theory and Research for the Sociology of Education*, edited by J. Richardson, 241–58. New York: Greenwood. https://www.marxists.org/reference/subject/philosophy/works/fr/bourdieu-forms-capital.htm.

(8)   Bourdieu, P. (1997). *Les Usages Sociaux de La Science : Pour Une Sociologie Clinique Du Champ Scientifique*. Paris: Éditions Quae.

(9)   Brunel University (2016). "The Revolution Will Not Be Televised (but It May Be Tweeted)." *Open Access Week Website*. http://openaccessweek.org/events/the-revolution-will-not-be-televised-but-it-may-be-tweeted.

(10) Budapest Open Access Initiative BOAI (2002). "Budapest Open Access Initiative." *Budapest Open Access Initiative Website*. http://www.budapestopenaccessinitiative.org/read.

(11) Camerer, C. (1988). "Gifts as Economic Signals and Social Symbols." *American Journal of Sociology* 94: S180–214.

(12) Enserink, M. (2016). "In Dramatic Statement, European Leaders Call for 'immediate' Open Access to All Scientific Papers by 2020." *Science,* May 27. doi:10.1126/science.aag0577.

(13) Gurstein, M. B. (2011). "Open Data: Empowering the Empowered or Effective Data Use for Everyone?" *First Monday* 16 (2). doi:10.5210/fm.v16i2.3316.

(14)	Herb, U. (2010). "Sociological Implications of Scientific Publishing: Open Access, Science, Society, Democracy and the Digital Divide." *First Monday* 15 (2). doi:10.5210/fm.v15i2.2599.

(15)	Herb, U. (2017). "Publishers of Journals Listed in the Directory of Open Access Journals (DOAJ)." *Zenodo.* doi:10.5281/ZENODO.838022.

(16)	Hilbert, M. (2014). "Technological Information Inequality as an Incessantly Moving Target: The Redistribution of Information and Communication Capacities between 1986 and 2010." *Journal of the Association for Information Science and Technology* 65 (4): 821–35. doi:10.1002/asi.23020.

(17)	Jahn, N., and M. Tullney. (2016). "A Study of Institutional Spending on Open Access Publication Fees in Germany." *PeerJ* 4 (August 9): e2323. doi:10.7717/peerj.2323.

(18)	Max-Planck-Society (2015). "Berlin 12 Homepage." *Max-Planck-Society Website.* https://openaccess. mpg.de/2128132/Berlin12.

(19)	Max-Planck-Society (2017). "Registration for B13 Open Access Conference is open." *Max-Planck-Digital-Library Website.* https://www.mpdl.mpg.de/en/about-us/news/408-registration-for-b13-open-access-conference-is-open.html.

(20)	Morrison, H. (2017). "From the Field: Elsevier as an Open Access Publisher." *The Charleston Advisor* 18 (3): 53–59. doi:10.5260/chara.18.3.53.

(21)	Murphy, K. (2016). "Should All Research Papers Be Free?" *New York Times*, March 12. https://www. nytimes.com/2016/03/13/opinion/sunday/should-all-research-papers-be-free.html.

(22)	OA2020–initiative for the large-scale transition to open access (2017). "13 Th Berlin Open Access Conference–Agenda." https://oa2020.org/wp-content/uploads/pdfs/B13-agenda.pdf.

(23)	OA2020–initiative for the large-scale transition to open access (2017). "B13 Conference–Building Capacity for the Transformation." *Open Access 2020 Website.* https://oa2020.org/b13-conference/.

(24)	Open Access Amsterdam (2016). "United Academics Open Access Conference." http://www. openaccess.amsterdam/.

(25)	Padula, D., and U. Herb (2016). "Questions Surrounding Affordable OA: Interview with Ulrich Herb." http://blog.scholasticahq.com/post/questions-surrounding-affordable-oa-ulrich-herb/.

(26)	Poynder, R. (2015). "The Open Access Movement Slips into Closed Mode." *Open and Shut?* http:// poynder.blogspot.de/2015/12/open-access-slips-into-closed-mode.html.

(27)	Prosser, D. (2003). "The Next Information Revolution—How Open Access Repositories and Journals Will Transform Scholarly Communications." *LIBER Quarterly* 14 (1). doi:10.18352/lq.7755.

(28)	Rankin, J. A., and S. G. Franklin (2004). "Open Access Publishing." *Emerging Infectious Diseases* 10 (7): 1352–53. doi:10.3201/eid1007.040122.

(29)	Weller, M. (2014). *The Battle For Open How Openness Won and Why It Doesn't Feel like Victory.* Ubiquity Press. doi:10.5334/bam.

Cooperative Futures: Technologies of the Common in the Collaborative Economy

Soenke Zehle

The creation of ambient media architectures brings machinic multiplicities into existence whose autonomy cannot be folded back easily into a politics of representation. In and of itself, this is nothing new—the autonomy of pollution particles or radiation waves has challenged attempts to regulate the consequences of their actions for a long time, giving rise to multiple bodies of thought, policy, and strategy in political ecology, systems design, and complexity governance. The interest in new forms of cooperation is driven largely by similar concerns, searching for ways of collaboration that allow a much higher degree of individual and collective self-determination to pursue shared concerns. Current debates on cooperativism take seriously the role of peer-to-peer logics in the shift from shared use to shared ownership, the power of computational infrastructures to scale local efforts beyond the boundaries of micropolitical solutions, and the need to affirm broader genealogies of the technological condition. Cooperativism research outlines a large horizon for action and analysis, exploring economic, social, and political strategies for an economy of shared ownership and collective self-organization. These social technologies of the common design the scene for cooperation.

Technologies of the Common

The question of the commons has been at the center of many reflections on economic, social, and cultural change. With a focus on sharing and the practices of commoning that create and sustain cultures of cooperation, accounts of collaborative economies have foregrounded the dynamics of enclosures and the practices of resistance to attempts to constrain shared use (6)(30)(31). In the context of data-driven platform economy business models and the "algorithmic governmentalities" governing their operations, these histories and traditions of co-ownership offer rich opportunities to reimagine, reframe and reorganize the way we live and work (41)(49). As contemporary peer-to-peer cultures experiment to redesign the cultural techniques of cooperation and collaboration, it seems all the more important to broaden the horizon of such experiments, beyond "sharing" as a technologically-driven practice and toward a much broader account of human-nonhuman sociality.[1]

Since "if 'commoning' has any meaning, it must be the production of ourselves as a common subject", there is a need to explore the role these dynamics of cooperation can play in shaping who we are and become, above and beyond the creation of fair economies (15). Framing these dynamics as "technologies of the common", this essay aims to contribute to such reflection. I started using the term "technologies of the common" a few years ago and have explored its analytical reach in different contexts (54)(55)(56)(57). While use of the term "technologies" seems to run counter to the need to challenge technological determinisms and technology-driven paradigms of cultural, economic, and social change, I think it is important to consider that the horizon of contemporary thought is indeed something like a "technological condition" whose implications we are just beginning to unfold (24). And whereas much reflection on commoning and the commons aims to counter the emphasis on technological innovation with an affirmation of the autonomy of social and other forms of non-technical innovation, I am convinced that it is on the terrain of the technological—of its infrastructural dynamics as well as the material conditions of possibility that structure its effects—that the scope of this autonomy will be determined. The stakes are high, and whether research is driven by the analysis of machinic socialities in "subjective economies" or curious explorations of emerging "cognitive assemblages" that cut across

---

1 Speaking of which (i.e. human sociality), the question of what it means (in an experience economy organized around attention rather than the visible) to hear, to heed, as a way to listen "to the language of things" (Walter Benjamin / Hito Steyerl) rather than to obey the injunctions to individuate (the resonances of a reading of Louis Althusser's account of interpellation as key dynamic of becoming-subject, the turning of a subject upon the calling of his/her name) runs through this essay as a sub-plot. Minor, but since the question continues to present itself as I write one might as well acknowledge such a presence.

human/non-human distinctions, it seems that we will have to think with and perhaps even from within the machine (22)(29).

The current debates on cooperativism take seriously the role of peer-to-peer logics in the shift from shared use to shared ownership, the power of computational infrastructures to scale local efforts beyond the boundaries of micropolitical solutions, and the need to affirm broader genealogies of the technological condition—beyond the "digital" in the narrow sense of post-WWII computational technology development or policy visions of the "digital society", both of which often end up reaffirming that our collective future is created in the former "Valley of Heart's Delight" on the US West Coast (32). As important, intriguing, and instructive as the interventions created in this influential, initially mainly publicly-funded research hub are, it makes little sense to accept these mythologies of innovation as new master narrative or policy template. A focus on cooperation brings a much wider and indeed transcultural archive of strategies into view, both of commons-based resource governance and the cosmopolitical visions informing them.[2]

Ours to Hack and Own

Across a series of events and publications, the ideas of a "platform cooperativism" and an "open cooperativism" have been developed to combine the dynamics of peer production with the cultural, economic and social concerns of cooperativism. Current cooperativism research outlines a large horizon for action and analysis, exploring economic, social, and political strategies for an economy of shared ownership and collective self-organization (4)(46). The ethos of cooperativism shapes not only the organization of collaborative research and development, but the design of technical systems to address the growing interest in transparent and trusted data-driven digital economy models. Integrating open technologies and the

---

2   The late Stuart Hall spoke of the "will to connect", a phrase that for me beautifully captures the excitement of explicit commitments to cooperation, but also directly link them to questions of power (as in a "will to power", growth, or greater intensity) such commitments imply. Maybe this is, at bottom, a Spinozist view of the world, but most importantly here is that the archive of such modes of cooperation must comprise more than another collection of toolkits for social change and include such "wills to connect". Otherwise, even a well-meaning cooperativism will find it difficult to open up the analytical and political horizon of cooperative social technologies and cultural techniques. I use the term "cosmopolitical" in the sense of Isabelle Stenger's "cosmopolitical proposal", see below.

concerns of open societies, cooperativism exemplifies the dynamics of open and social innovation in the collaborative economy and combines techno-logical and non-technological innovation in unique ways. Across this lit-erature of research and practice, I find the following themes particularly relevant in facilitating reflection on technologies of the common:

Histories of Non-Technical Innovation: If we think of innovation in technological terms, the historical horizon of our current moment is mainly defined by the horizon of technological innovation. Histories of Silicon Valley show that the combination of massive post-war public invest-ment in research infrastructures and the cultural legacies of a "hippie mod-ernism" does not provide a template that can easily be reproduced simply by bringing venture capital together with start-up incubators and acceler-ators (8). Cooperativism offers organizational alternatives to the corpora-tion and reframe the all-too-brief histories of the digital society, beyond innovation ecosystems clustered around the oligopolies of the platform economy, but also beyond accounts that feature mainly male engineering heroes.[3] This is crucial. While the rediscovery of historical practices of experimentation and sharing in the name of "social innovation" is wel-come, the consequences of digital transformation require more than another wave of innovations to repair whatever collateral damage techno-logical determinism might cause. Cooperativism shifts the focus of debate from the fascination with expert-driven "innovation" toward the broader cultural contestation of how we actually organize life and work in a future of shared experience.

Members not Markets: Cooperativism is not a communism-to-come but the pragmatism of coupling traditions and technologies in new assem-blages. Many innovations are simply useful rather than disruptive and therefore "make better communities than commodities" (43). In such cases, the logic of venture capital pushing start-ups towards buyouts and IPOs makes less sense than cooperative ownership models designed to meet the needs of members rather than stock markets.[4] At the same time,

---

3   Online reviewers of a recent history of the "digital universe" (13), for example, largely focus on the relationship between John von Neumann and ENIAC pioneers-turned-house-hold-names J. Presper Eckert and John W. Mauchly. Rarely mentioned is that "Kay McNulty, Betty Jennings, Betty Snyder, Marlyn Meltzer, Fran Bilas, and Ruth Lichterman were the original programmers of the first American electronic computer, and they have tra-ditionally been little more than a footnote in the history of the ENIAC"; see (19). Also see (5)(20)(23)(26)(47).

4   See http://platformcoop.net/resources/directory for a directory of platform cooperatives.

it makes more sense to speak of "member markets" rather than "members not markets"—not all markets are the same, and to limit the horizon of cooperativisms to "non-market" economies might mean many a missed opportunity for the creation of larger-scale alternatives.[5]

Use as Work: Behind this pragmatism lies an analysis of the so-called "sharing economy" that approaches use as a form of work (34)(45). In the data-driven business models of the platform economy, users generate the data, the interfaces of free services operate as devices of capture, and comprehensive analytics apparatuses trace user movements across the web as well as the site in question.[6] Because cooperatives have roots in worker-driven forms of labor organization, the logic of cooperatives can bring hackers and workers together.[7] These participation-as-free-labor analyses are part of a broader set of concerns regarding the rules of the attention economy. If users want to use ad blockers to modify their experience of use to screen out the crass commercialism of online ads, for example, companies contend that users are undermining their business model and weaken print-to-online transition strategies. What they don't say is that they want to collect the data without telling users what they do with it (36)(39)(44). While we are lightyears away from data transparency, new forms of activism are already reclaiming the terrain of the attention economy—from algorithmic accountability journalism to concerted efforts to blacklist (and thereby defund) online sites.[8] How users can get involved in the negotiation of the trade-offs of the benefits of cloud-based data-driven approaches

---

5   Blockchain-driven cooperatives like the streaming platform resonate.is watch the shift from IPOs to ICOs (Initial Coin Offerings) closely and expect major changes in the way (cooperative) platform projects will be financed.

6   The ability to trace users across the web prompted a key shift in Facebook's relationship to online advertising (28), also see (33).

7   In his analysis of "vectoralism", McKenzie Wark writes: "What the vectoralist firm owns and controls is brands, patents copyrights, and trademarks, or it controls the networks, clouds, and infrastructures, along which such information might move. The rise of the so-called sharing economy is really just a logical extension of this contracting out of actual material services and labor by firms that control unequal flows of information. ... The significance of platform cooperativism is that it is a movement that can place itself at the nexus of the interests and experiences of both workers and hackers. Why not use the specific skills hackers have to create the means of organizing information, but use it to create quite other ways of organizing labor? Cooperatives have a long history in the labor movement; indeed, in their origins, they looked back to forms of peasant self-organization of the commons" (53).

8   See https://www.facebook.com/slpnggiants for a campaign to defund breitbart.com.

(customization, efficiency, etc.) and act on their growing awareness of the risks of such automated data collection by companies and state agencies is highly relevant to cooperative "systems design" strategies focusing on co-owned, federated infrastructures.

Urban Commons. While commons are often associated with the shared ownership and use of natural resources, the question of commoning in urban environments has been taken up by a diverse group of actors interested in "the city as a commons" (17). Many economic activities across the collaborative economy are decentralized, but the processes of social innovation are often rooted in the dynamics of urban life and closely linked to urban governance strategies. Cities can do much to facilitate these forms of innovation, and municipal actors in 'rebel' cities like Barcelona are demonstrating the potential of commons-based and peer-to-peer approaches in the design of comprehensive urban governance strategies (2)(18)(21)(42). The sharing of such models across city networks is a form of political organization that tends to attract less attention than the activities of states or supranational organizations. But since cities offer rich opportunities to make alternative futures tangible, cooperativism has set its sights on a networked municipalism that allows ordinary people to become active across borders.

New Alliances. Easier said than done, the experiences of DIY and maker efforts suggest that it is through hands-on experimentation and collaborative prototyping that such broadening of debates occurs. Here, new alliances might be on the horizon—SMEs unsure about how to approach and invest in the large number of technologies loosely grouped in "industry 4.0" or "industrial internet" policy frameworks welcome greater opportunities to explore these technologies before they invest. And their approach to the dynamics of the platform economy might have more in common with (un)civil society organizations than large corporate players.[9]

Another Europe. While radical municipalism has a long history in the US, the new meso-politics—neither individual nor national but a politics of and across the in-between-structures—of cross-city cooperation is of particular relevance to Europe. While it is hard to believe that we still have to make the case for cross-border cooperation, in the face of a

---

9   In Europe, SME-focused "digital innovation hubs" (https://ec.europa.eu/digital-single-market/en/digital-innovation-hubs) exist alongside grassroots-driven "digital social innovation" sites (http://digitalsocial.eu)

well-coordinated resurgence of cultural and economic nationalisms it is important to document that another Europe already exists. Much of this other Europe revolves around the collaborative economy.[10] Cooperatives are currently at the heart of a key innovation debate around the platform economies that have become the signature infrastructures of a new generation of services in the collaborative economy.[11] Following the 2012 UN Year of Cooperatives, the 2013 establishment of the EU Working Group on Cooperatives, and the 2016 Bratislava Declaration, cooperatives have been widely recognized as a practice of cultural, economic and social collaboration that is also a core dynamic of non-technological innovation.[12] The cooperative economy ethos is much older than the digital economy, yet analyses of current innovation models frequently fail to take the wide spectrum of historical models into account. This is where a new agenda of cooperativism research can make key contributions.[13]

Social Technologies. The traditions of small-scale "common-pool resource management" practices (Elinor Ostrom) have already featured prominently in collaborative economies.[14] Such approaches are currently being complemented by the transfer of the logic of blockchains, the distributed ledger system powering virtual currencies, to other forms of organization (16). What makes the new generation of cooperative ventures so interesting is that they address the question of scale (4)(40). Based on fundamental assumptions about how trust works and can be encoded, blockchains are already expected to be involved in a wide range of applications (51). As with all technological innovations, there is nothing "inherently emancipatory" about the blockchain (10). The question is, again as always, how design and use are framed, how we balance the exercise of individual

---

10   https://ec.europa.eu/growth/single-market/services/collaborative-economy_en

11   The legal statute of a European Cooperative Society (https://ec.europa.eu/growth/sectors/social-economy/cooperatives/european-cooperative-society_en, https://coopseurope.coop/policy-topic/regulatory-framework-cooperatives) provides the idea of a Europe from below with a figure of law. Also see (11).

12   As of 2015, cooperatives reached "a total annual turnover of 1,005 billion Euros—more than the GDP of Finland, Denmark, Norway and Sweden combined" (12)(14).

13   In 2016, Germany's tradition of cooperatives was added to UNESCO's Intangible Cultural Heritage list.

14   In smaller-scale collaborations, the gains of cooperation outweigh the benefits of (narrowly) self-interested action. See (38). Research by Ostrom and her colleagues has been taken up widely across alternative, solidarity economy efforts and continues to serve as shared point of reference. See, for example, (7).

and social rights? Will blockchain-based cooperatives thrive only in gentrified neighborhoods blessed with broadband access and generous technological literacy resources? If we get involved in technology design debates over "What kinds of subjectivity do we want to algorithmically inscribe into our systems?,", these debates need to involve a much wider array of people from beyond the confines of coder cultures (35)(9).

Cooperative Machinisms. It seems that just as we are about to transfer the logics of peer-to-peer communication across the fields of cultural, economic, and social activity, the internet itself is changing: "the telecoms industry has evolved from a public peer-to-peer service—where people had the right to access telecommunications—to a pack of content delivery networks where the rules are written by a handful of content owners, ignoring any concept of national sovereignty" (25). What is more, this shift occurs in the context of efforts to "democratize" artificial intelligence without a corresponding democratization of the (oligopolistic) infrastructures framing such engagement.[15] So while it is exciting to follow the invitation to "democratize" cloud-based AI systems, there is no need to lose sight of the broader question of who owns the information spaces within which we are expected to conduct our affairs.[16] In the context of cooperativism, these trends—engaging with infrastructural changes and the potential of machine learning—are already coming together.[17] Above and beyond individual acts of sharing or strategies of resource management, cooperation

---

15   See, for example, Google's Chief Scientist of Google Cloud and Machine Learning Fei Fei Li on the "democratization" of AI in her keynote for Google Cloud Next 2017, https://www.youtube.com/watch?v=Rgqgdddl018; http://ai-4-all.org/; or Elemental Cognition (https://www.elementalcognition.com), founded by former IBM Watson-engineer David Ferruci.

16   Less obvious perhaps than questions of internet governance, cooperatives are also involved in the design of co-owned energy systems to offer alternatives to top-down infrastructuralism approaches. See Cooperatives Europe, RESCoop.Eu, Enercoop, The Co-operative Energy, SomEnergia, ICLEI, Climate Alliance, Friends of the Earth Europe and Client Earth, "Joint Reaction to the Energy Union Package of the EU Commission of 25/03/2015", (25.03.2015), https://coopseurope.coop/sites/default/files/Reaction%20to%20Energy%20Union%20Package%20Communication%5B1%5D.pdf. For "a snapshot of the European attempt to turn infrastructural connectivity into a new form of collectivity", see (37). See the European Federation of Renewable Energy Cooperatives (https://rescoop.eu/), also see https://www.indigoadvisorygroup.com/blockchain for a collection of blockchain-based use cases across the industry.

17   See the collaboration tool http://pol.is that incorporates machine learning to scale democratic deliberations, already used in Taiwan's "vTaiwan" collaborative policy development project. On vTaiwan, see (3)(48)

can be understood as a cultural technique with a long cross-cultural history. As we engage in the algorithmicization of our core cultural techniques, this affects cooperation as well. A better understanding of how we share, of what sharing is, not just of how we engage in acts of sharing across solidarity economies, but of who we become as we organize ourselves as shared selves across the infrastructures of distribution (social media, data-driven economies of work).

Designing a Cosmopolitics of the Common

The creation of ambient media architectures brings machinic multiplicities into existence whose autonomy cannot be folded back easily into a politics of representation. In and of itself, this is nothing new—the autonomy of pollution particles or radiation waves has challenged attempts to regulate the consequences of their actions for a long time, giving rise to multiple bodies of thought, policy, and strategy in political ecology, systems design, and complexity governance. The interest in new forms of cooperation is driven largely by similar concerns, searching for ways of collaboration that allow a much higher degree of individual and collective self-determination to pursue shared concerns. Because such sharing involves—literally, folds into a space of shared experience—human and non-human actors, it calls for(th) a cosmopolitics of the common rather than the politics of representation we already know how to organize. "The fundamental problem we have is that technologies are only as good as their makers. There is mounting evidence that machine-learning algorithms, like all previous technologies, bear the imprint of their designers and culture. ... Making the politics of algorithms visible, explicit and accountable may turn out to be even more difficult than calling, say, lawyers to account. ... The point of these scenario-building exercises is precisely to authorize the participation of a broad range of relevant actors typically excluded from processes of deliberation about the future" (52).

The cosmopolitical opens up an analytical horizon beyond that of "smart" citizenship and of human voices articulating already-defined interests. In Isabelle Stenger's vision, "the proposal is open to misunderstanding, liable to the Kantian temptation of inferring that politics should aim at allowing a 'cosmos', a 'good common world' to exist—while the idea is precisely to slow down the construction of this common world, to create a space for hesitation regarding what it means to say 'good'" (50).

Whatever views of the good society motivate the new wave of cooperativisms, the concern of a cosmopolitics of the common is not to adjudicate the differences of the worlds imagined. Quite the contrary: "The cosmopolitical proposal is incapable of giving a 'good' definition of the procedures that allow us to achieve the 'good' definition of a 'good' common world" (50). To heed the cosmopolitical proposal is "a matter of imbuing political voices with the feeling that they do not master the situation they discuss, that the political arena is peopled with shadows of that which does not have, cannot have or does not want to have a political voice ... inventing the way in which 'politics', our signature, could proceed, construct its legitimate reasons, 'in the presence of' that which remains deaf to this legitimacy: that is the cosmopolitical proposal".[18] The common in such a cosmopolitics is not the transcendent universal ground of a politics of rights in which autonomous subjects create institutions to govern their affairs, but refers to the actuality of material interdependencies that cut across culture and nature, subjects and objects, selves and others.

A cosmopolitics of the common is a matter of design: "How to design the political scene in a way that actively protects it from the fiction that 'humans of good will decide in the name of the general interest'? ... But also how to design it in such a way that collective thinking has to proceed 'in the presence of' those who would otherwise be likely to be disqualified as having idiotically nothing to propose, hindering the emergent 'common account'? Designing a scene is an art of staging" (50). So how do we design the scene for cooperation? The debate has already begun, quickly gathering into archives to facilitate sharing. Opportunities to reflect on the design of the socio-technological worlds that shape life and labor do not simply arise. They have to be created. To join does not require much, so let us get involved on a broad scale.

It certainly does not require expertise, not even a commitment to any kind of politics. Stenger's invitation to "laugh not at theories but at the authority associated with them" offers a point of departure that is literally always already shared—as is the nature of laughter as a social gesture, and as is the nature of the common (1). To laugh at the presumptuousness of (largely humorless) theories of the digital society, for example, is a possible first step towards a practice of commoning, which begins not

---

18 This also offers a way to attend to the call to create "cultures of failure" that is so central to innovation strategies (fail often, fail early, fail cheaply) by hearing it differently, as "the building up of an active memory of the way solutions that we might have considered promising turn out to be failures, deformations or perversions" (50).

with new social technologies or the pragmatism of political solutions but with denying master narratives whatever authority is attributed to them and allows them to operate as a symbolic fictions lending coherence, legitimacy, and power to our worlds (58).[19]

19   One could continue this, of course—constituted power is tragic, constituting power is comic etc. And if the humour and irony at the heart of human "resilience" (a popular meme in "smart" societies visions that comprehend resilience mainly as a feature of systems rather than a register of sociality) can ridicule the short-sightedness of technological determinisms, the acknowledgment that to think is to resist (Hannah Arendt, Gilles Deleuze) is only a small step away.

References

(1)   De Angelis, M., and D. Harvie (2015). The common. In *The Routledge Companion to Alternative Organization*, edited by Parker, M., et al., 280–94. New York: Routledge.

(2)   Barcelona en Comú (2016). *How to Win Back the City En Comé: Guide to Building a Citizen Municipal Platform*. Barcelona. https://barcelonaencomu.cat/sites/default/files/win-the-city-guide.pdf

(3)   Barry, L. (2016). *vTaiwan: Public Participation Methods on the Cyberpunk Frontier of Democracy*. Civicist August 11, 2016. https://civichall.org/civicist/vtaiwan-democracy-frontier/

(4)   Bauwens, M., and V. Kostakis (2014). From the Communism of Capital to Capital for the Commons: Towards an Open Co-operativism. *Triple C* 12(1).

(5)   Beyer, K.W. (2009). *Grace Hopper and the Invention of the Information Age*. Cambridge: MIT Press.

(6)   Bollier, D. (2014). *Think Like a Commoner*. Gabriola Island, BC: New Society Publishers.

(7)   Bollier, D., and S. Helfrich, eds. (2012). *The Wealth of the Commons: A World beyond Market and State*. Amherst and Florence, MA: Levellers Press.

(8)   Castillo, G., E. Choi, and A. Clarke (2015). *Hippie Modernism: The Struggle for Utopia*. Minnesota: Walker Art Center.

(9)   Catlow, R., and P. Gomes (2016). *The Blockchain: Change everything forever*. http://ruthcatlow. net/?works=the-blockchain-change-everything-forever

(10)   Chakrabarti, U.K. (2015). *From Bearer Bonds to the Blockchain: Artistic Perspectives on Digital Money*. eatthehipster (01/05/2015), https://eatthehipster.org/2016/05/01/blockchains-potential-in-the-arts/

(11)   Como, E., et al. (nd). *Cooperative Platforms in a European Economy: An Exploratory Study*. https:// coopseurope.coop/sites/default/files/Paper_Cooperatives%20Collab%20Economy__0.pdf

(12)   Cooperative Europe (2015). *The Power of Cooperation: Cooperatives Europe Key Statistics 2015*. https://coopseurope.coop/sites/default/files/The%20power%20of%20Cooperation%20-%20 Cooperatives%20Europe%20key%20statistics%202015.pdf

(13)   Dyson, G. (2012). *Turing Cathedral: The Origins of the Digital Universe*. New York: Vintage.

(14)   Esser, F.C., et al. (2016). *The European Collaborative Economy: A research agenda for policy support*. JRC Science Policy Report. EUR 28190 EN; 10.2760/755793. Luxembourg: Publications Office of the European Union.

(15)   Federici, S. (2012). *Revolution at Point Zero: Housework, Reproduction, and Feminist Struggle*. Oakland, CA: PM Press.

(16)   De Filippi, P. (2016). The Interplay between Decentralization and Privacy: The Case of Blockchain Technologies. *Journal of Peer Production* 9.

(17)   Foster, S., and C. Iaione (2016). The City as a Commons. *Yale Law and Policy Review* 34, 281–349.

(18)   Fuster Morell, M. (2017). *Barcelona as a Case Study on Urban Policy for Platform Cooperativism.* p2pfoundation. https://blog.p2pfoundation.net/mayo-fuster-morell-barcelona-as-a-case-study-on-urban-policy-for-platform-cooperativism/2017/02/23

(19)   Gidalevitz, Y. (2017). *The surprisingly unknown history of women in computing.* Invision. http://blog.invisionapp.com/history-of-women-computing/

(20)   Green V. (2001). *Race on the Line: Gender, Labor, and Technology in the Bell System, 1880–1980.* Durham: Duke University Press.

(21)   Harvey, D. (2012). *Rebel Cities: From the Right to the City to the Urban Revolution.* London and New York: Verso.

(22)   Hayles, N.K. (2017). *Unthought: The Power of the Cognitive Nonconscious.* Chicago: Chicago University Press.

(23)   Hicks, M. (2017). *Programmed Inequality: How Britain Discarded Women Technologists and Lost Its Edge in Computing.* Cambridge, MA: MIT Press.

(24)   Hoerl, E. (2015). The Technological Condition (trans. A. Enns). *Parrhesia* 22, 1–15.

(25)   Huston, G. (2017). *The Internet's Gilded Age.* Circle ID. http://www.circleid.com/posts/20170311_the_internet_gilded_age/

(26)   Isaacson, W. (2014). *The Innovators: How a Group of Hackers, Geniuses, and Geeks Created the Digital Revolution.* New York: Simon & Schuster.

(27)   Kostakis, V., A. Pazaitis, and M. Bauwens (2016). Digital economy and the rise of open cooperativism: The case of the Enspiral Network. *Working Papers in Technology Governance and Economic Dynamics no. 68.*

(28)   Langley, H. (2016). *Facebook will now track you across the web, even if you don't have an account.* Techradar. http://www.techradar.com/news/internet/facebook-will-now-track-you-across-the-web-even-if-you-don-t-have-an-account-1322278

(29)   Lazzarato, M. (2013). *Signs and Machines: Capitalism and the Production of Subjectivity.* Cambridge MA: MIT Press.

(30)   Linebaugh, P; (2008). *The Magna Carta Manifesto: Liberties and Commons for All.* Berkeley, CA: University of California Press.

(31)   Linebaugh, P. (2014). *Stop, Thief! The Commons, Enclosures, and Resistance.* Oakland, CA: PM Press.

(32)   Malone, M.S. (2002). *The Valley of Heart's Delight: A Silicon Valley Notebook 1963—2001.* London: Wiley.

(33)   Martinez, A.G. (2016). *Chaos Monkeys: Obscene Fortune and Random Failure in Silicon Valley.* New York: Harper.

(34)   Maxwell, R., ed. (2015). *The Routledge Companion to Labor and Media.* New York: Routledge.

(35)   O'Dywer, R. (2015). *The Revolution will (not) be decentralised: Blockchains.* Commons Transition. http://commonstransition.org/the-revolution-will-not-be-decentralised-blockchains/

(36)   O'Neil, C. (2016). *Weapons of Math Destruction: How Big Data Increases Inequality and Threatens Democracy.* New York: Crown.

(37) Opitz, S., and U. Tellmann (2016). Europe's Materialism: Infrastructures and Political Space. *Limn Magazine* 7.

(38) Ostrom, E., et al. (2012). *The Future of the Commons: Beyond Market Failure and Government Regulations*. Institute of Economic Affairs: Occasional Papers. London: IEA.

(39) Pasquale, F. (2015). *Black Box Society: The Secret Algorithms That Control Money and Information*. Cambridge MA: Harvard University Press.

(40) Pazaitis, A., P. De Filippi, and V. Kostakis (2017). Blockchain and Value Systems in the Sharing Economy: The Illustrative Case of Backfeed. *Working Papers in Technology Governance and Economic Dynamics* no. 73.

(41) Rouvroy, A. (2015). *Of Data and Men. Fundamental Rights and Freedoms in a World of Big Data*. Strasbourg: Council of Europe.

(42) Russell, B., and O. Reyes (2017). *Eight lessons from Barcelona en Comú on how to Take Back Control*. Open Democracy. https://www.opendemocracy.net/can-europe-make-it/oscar-reyes-bertie-russell/eight-lessons-from-barcelona-en-com-on-how-to-take-bac

(43) Schneider, N. (2017). *Users should be able to own the businesses they love instead of investors*. Quartz Ideas. https://qz.com/940876/sell-businesses-to-their-users-instead-of-investors/

(44) Schneier, B. (2015). *Data and Goliath: The Hidden Battles to Collect Your Data and Control Your World*. New York: W. W. Norton.

(45) Scholz, T. ed. (2012). *Digital Labor: The Internet as Playground and Factory*. New York: Routledge.

(46) Scholz, T., and N. Schneider, eds. (2017). *Ours to Hack and to Own. The Rise of Platform Cooperativism, a New Vision for the Future of Work and a Fairer Internet*. New York City: OR Books.

(47) Shetterly, M.L. (2016). *Hidden Figures: The American Dream and the Untold Story of the Black Women Mathematicians Who Helped Win the Space Race*. New York: William Morrow.

(48) Simon, J., et al. (2017). *Digital Democracy: The tools transforming political engagement*. London: NESTA.

(49) Srnicek, N. (2017). *Platform Capitalism*. Cambridge: Polity Press.

(50) Stengers, I. (2005). The Cosmopolitical Proposal. In: *Making Things Public*, edited by Latour, B., and P. Weibel, 994–1003. Cambridge MA: MIT Press.

(51) Valpicelli, G. (2016). Beyond bitcoin. Your life is destined for the blockchain. *WIRED* (June 8).

(52) Wacjman, J. (2017). Automation: Is it really different this time? *Britisch Journal of Sociology* 68 (1), 119–127.

(53) McKenzie Wark, "Worse than Capitalism", in Scholz and Schneider , 45–46.

(54) Zehle, S. (2006). Technologies du commun: nouvelles subjectivités et société civile globale. *Vacarme* 34, 90–5;

(55) Zehle, S. (2007). Technologies of the Common: Toward an Ethics of Collaborative Constitutio. In *Éthique et droits de l'homme dans la société de l'information = Ethics and human rights in the information society: Proceedings, Synthesis and Recommendations / European Regional Conference, 13 -14 September 2007, Strasbourg, France. Paris: Commission Nationale Française pour l'UNESCO*.

(56) Zehle, S. (2014). Reclaiming the Ambient Commons: Strategies of Depletion Design in the Subjective Economy. *International Review of Information Ethics*, Special Issue: Ethics for the Internet of Things 22 (12), 31–42.

(57) Zehle, S. (2016). Common Conflicts, Imperial Imaginaries: Exploring the Becoming-Environmental of Media. In: *What Urban Media Art Can Do*, edited by Pop, S. et al., 420–28. Stuttgart: avedition.

(58) Žižek, S. (2014). *Žižek›s Jokes (Did you hear the one about Hegel and negation?)*, edited by Mortensen, A. Cambridge MA: MIT Press.

Part Two:
North/South

_____

The Contribution of the Global South to Open Access

Hélène Prost and Joachim Schöpfel

What is the actual contribution of the Global South to the open access move-ment? Do open repositories and academic journals in open access change the situation of unequal scientific production? The question is quite simple but the answer isn't, and this is true for three reasons. Monitoring open access is still a problem, and despite useful and efficient directories and discovery tools, nobody can provide reliable information on the content in open access. Also, because of their bias in favour of great research countries like the United States, UK, Germany or France, emerging countries and the Global South in general are less visible and underrepresented in these tools. Finally, the very term of Global South is fuzzy; what exactly is the Global South? The following chapter tries to provide some empirical clarity to support a better understanding of the situation.

A map of global inequality

The map of global knowledge production is a map of global inequality (Figure 1). The production and exchange of scientific papers is domi-nated by some major research-intensive countries, nearly all located in the Northern hemisphere. Is this map destiny? Laura Czerniewicz—who ana-lysed the underlying, inequitable global power dynamics—is convinced that this situation must be confronted and can be challenged. She also suggests that "the open access movement needs to broaden its focus from access to knowledge to full participation in knowledge creation and in scholarly communication"[1].

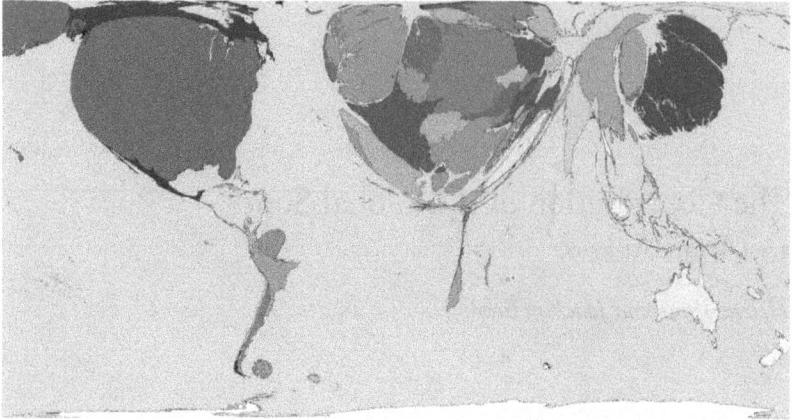

Figure 1. Scientific papers published in 2001[2]

In light of Czerniewicz's critique, we ask: What is the actual contribution of the Global South to the open access movement? Do open repositories and academic journals in open access change the current situation of unequal scientific production?

The question is quite simple but the answer isn't, and this for three reasons. Monitoring open access is still a problem, and despite useful and efficient directories and discovery tools, nobody can provide reliable information on the number of articles in open repositories or open access journals, and even less in other categories of scientific literature such as grey literature. Also, because of their bias in favor of great research countries like the United States, UK, Germany or France, emerging countries and the Global South in general are less visible and underrepresented in these tools. Finally, the very term Global South is fuzzy; what exactly is the Global South? Where are the frontiers with the North? Our selection of 101 countries (Figure 2) is based on economic data from the World Trade Organization and the World Bank Group (1), and we exploited the country

---

1 http://www.socialsciencespace.com/2015/07/this-global-science-map-is-not-destiny/

2 © Copyright Worldmapper.org / Sasi Group (University of Sheffield) and Mark Newman (University of Michigan)

information provided by Elsevier's scientometric database (Scopus), the Directory of Open Access Repositories (OpenDOAR), the Directory of Open Access Journals (DOAJ), the Bielefeld Academic Search Engine (BASE), and the Registry for Research Data Repositories (re3data) run by DataCite. All data were gathered in March 2017.

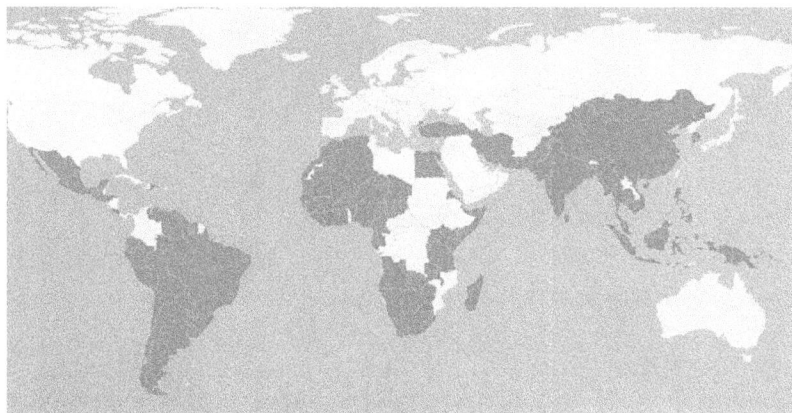

Figure 2. The Global South countries in our study (N=101)

Because of the data sources (WTO and World Bank Group), a couple of smaller and some major countries were missing in the original list of 99 countries, above all and especially Algeria and Iran. Both were added to our sample by hand.

Global and open production

In March 2017, the Bielefeld Academic Search Engine (BASE) indexed more than 107m documents from 5,300 providers. 43m items are freely available on internet, that is, indexed as open access. Only 2.6m or 6% of these open access items are provided from the Global South countries, and from three countries in particular namely Brazil, India and Indonesia, which together represent nearly 60% of the Global South items in open access (Figure 3). On the other hand, 55 Global South countries are not indexed in BASE and their open access publications remain virtually non-existent and invisible.

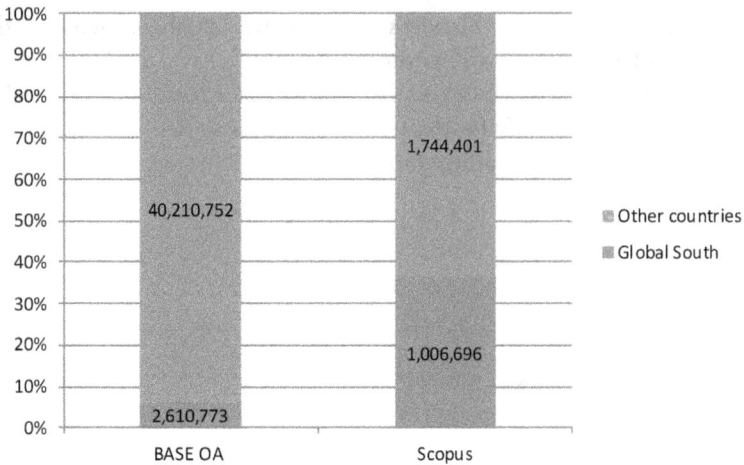

Figure 3. Global South academic output in BASE (open access, cumulative) and Scopus (2016)

Compared to the Scopus database, this percentage seems far too low. One third (1m) of the 2016 scientific output referenced by Scopus comes from the Global South, and this output is dominated by China, India and South Korea, which represent 68% of the whole Global South while one third of the Global South countries are without any or with a very low output (<100). This discrepancy between 6% (BASE open access) and 37% (Scopus 2016) is due in large part to the nearly complete absence of Chinese open access papers in BASE. But it may also be that the European, North American or Japanese open access publications are better represented because of the rapid increase of gold open access with article processing charges (APCs) in the Northern Hemisphere, which is so far without a counterpart in the Global South. When considering only the 2016 content of BASE, the share of the Global South increases to 10%, which appears to reflect a growing importance of their contribution to international science.

Open repositories

The BASE Figures provide information about the number of documents but do not tell us anything about projects, initiatives, or the gold vs. green road. What do we know about the green road, that is, open repositories? In March 2017, the OpenDOAR database contains 3,335 repositories of which 63% are hosted by organizations in Europe and North America.

Only 795 repositories (24%) are in the Global South, with a strong concentration in four countries—40% of the repositories from the Global South are in Brazil, India, Indonesia and Turkey.

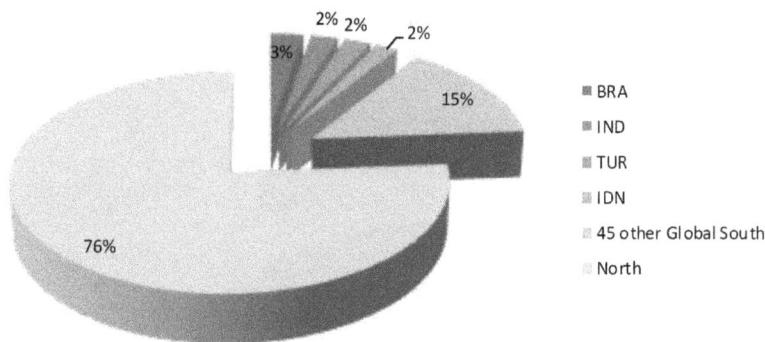

Figure 4. Open repositories in Global South countries

As for the other countries the major portions of these repositories are institutional repositories, launched and maintained by universities and research organisations. Many of them are run with the MIT DSpace software, a smaller part with Greenstone, developed and distributed in cooperation with UNESCO and the Human Info NGO, with EPrints (University of Southampton) and, especially in South America, with the SciELO platform. As for the content, the share of grey literature, in particular theses and dissertations, seems higher than in Europe or North America where published journal articles prevail.

Another difference is the relative importance of languages other than English—Spanish and Portuguese in South America; Chinese, Turkish, Indonesian, Korean and Arabic in Asia and the Middle East; and French and Arabic in Africa. However, English remains by far the most important language in open repositories and the Lingua Franca of scientific research also in the Global South.

Open access journals

What do we know about the gold road, that is, open access journal publishing in the Global South? First of all, one third of the open access journals are edited in Global South countries. In March 2017 the DOAJ directory registers 9,400 titles, and 3,685 are edited in one of the 101 countries of our list (39%). Again, a small number of countries dominate the market, that is, Brazil (10%), Egypt (6%) and Indonesia (6%), followed by Iran, India and Turkey (each with 3%) (Figure 5). However, recent studies reveal that these figures are probably too low and that especially Chinese institutions and publishers edit many more journals in open access than the DOAJ shows (7).

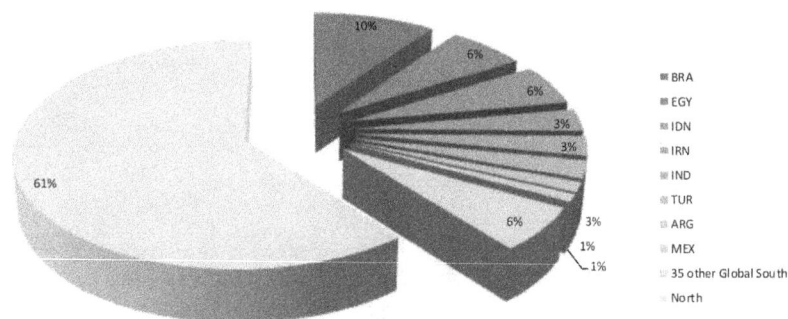

Figure 5. Open access journals in Global South countries

Another look reveals different situations—while in Brazil, Indonesia and Iran most of the journals are published by institutions or learned societies and do not apply APCs, in Egypt commercial open access publishing with APCs prevails (71%). In India and Turkey, most journals do not provide information about their APC policy. The situation was quite different until 2016, when DOAJ removed 3,000 titles and applied a new and more selective policy. Many of the removed titles were edited by Indian publishers suspected of predatory publishing (2).

Data repositories

In the context of open science, data repositories are becoming increasingly important. Nearly all—94%—of the data repositories listed in the re3data directory are open and can provide open access to research data.

An important caveat is that not all of the data in open repositories are freely available due to privacy concerns, intellectual property issues and so forth. Re3data is the most comprehensive directory of data repositories worldwide but because of its funding background it is (still) dominated by four countries, the US, Germany, the UK and Canada which represent 81% of the repositories and 71% of all registered institutions (5). For many other countries data are simply not available, especially in Africa, South America, the Middle East and South-East Asia.

So it may be even more surprising that among the 2,579 data repositories indexed by re3data we are able to find 113 repositories from the Global South (4%); 109 of them are open. These repositories are hosted in 23 countries, led by China (32), India (30) and Mexico (11). Again, the small number can probably best be explained by the delayed uptake of the re3data initiative outside Western Europe and North America; however, lack of data infrastructures and security in many countries surely contribute to the situation.

Correlations

Our Figures show so far a contrasted landscape, with large disparities between the different variables (Figure 6).

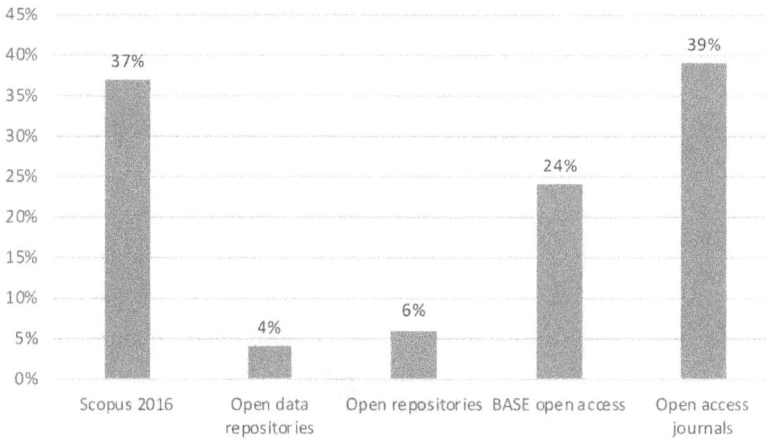

Figure 6. Synthesis of open access in the Global South (in %, N=101)

While these 101 countries count for 37% of the 2016 scientific publications indexed by the Scopus database, they represent only 6% of the open repositories, 24% of the open items retrieved by BASE and 39% of all open access journals. This may be surprising at first sight, as the prevalence of open access journals seems to challenge the idea that "gold is good for the rich and green for the rest". But for newcomers, open access journals are an easy way to enter the academic information market, as they respond to the increasing demand for quick and easy dissemination of research results. For institutions and learned societies, on the other hand, the gold road provides a win-win option to increase impact and visibility and to guarantee the usual level of quality assurance via peer review.

The number of repositories and open access journals are correlated (Figure 7) with Pearson's r=.70. This correlation increases after elimination of "atypical" countries with high numbers of open access journals like Brazil, Egypt, Indonesia or India.

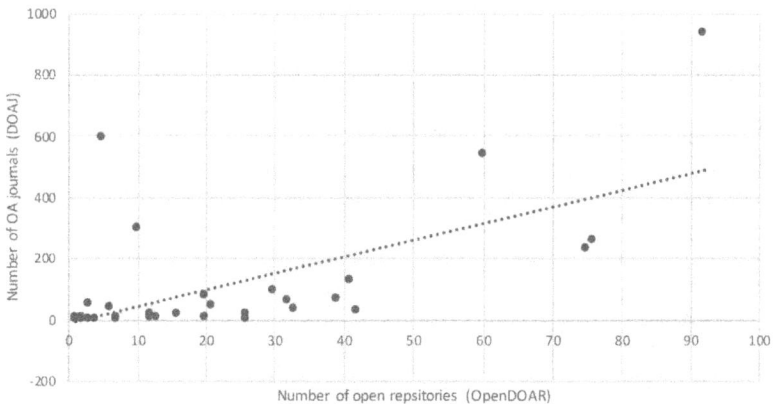

Figure 7. Correlation between open repositories and open access journals (N=101)

In other words: except for those mentioned above, most countries do not choose between green and gold strategies but develop both, at a relatively low level and often based on institutional initiatives, mostly university repositories and journals. Both variables, repositories and journals, are highly correlated with the open access content retrieved by BASE (r>.80) but there is no significant link with the scientific production indexed by Scopus (r=.20-.40) which appears, at least for the Global South countries, to reflect primarily traditional journal publishing. Perhaps more surprising is the high correlation between the Scopus output and the number of open data repositories (r=.88) which may indicate that countries with visible academic output invest more in data infrastructures and more consciously attend to their visibility and impact.

Geographical areas

	1	2	3	4	5
Latin America & Caribbean	28	123523	1402789	294	1380
South Asia	5	157739	485635	104	349
East Asia & Pacific	22	609198	367174	175	724
Europe & Central Asia	1	43593	244277	75	235
Middle East & North Africa	6	87014	51582	31	917
Sub-Saharan Africa	39	41403	46296	116	80

Figure 8. Open access in different areas of the Global South (1. countries, 2. publications 2016, 3. documents in open access, 4. repositories, 5. journals)

The same variety of situations can be observed between the different areas of the Global South (Figure 8). Latin America & Caribbean is by far the most important open access area, with the greatest number of documents in open access (BASE), of repositories (OpenDOAR) and of journals (DOAJ) and with two leading countries, Argentina and especially Brazil.

South Asia ranks second regarding the number of documents freely available via BASE, but only fourth in terms of repositories and journals. The main country in this area is India, with large repositories and relatively few referenced journals for the reason mentioned above (predatory publishing).

East Asia, the first area of the Global South in terms of academic output, ranks third with regards to open access, with several hundreds of journals and 175 repositories. Surprisingly, the most important open access country today is neither China nor South Korea but Indonesia which counts 541 open access journals (mostly university presses) and 60 institutional repositories.

Figure 9 illustrates these differences and shows in particular the specific place of the other three areas:

Middle East & North Africa: Many journals but few open repositories and few documents in open access. The most important country is Egypt, which hosts the Hindawi Publishing Corporation, a leader on the market of commercial open access journals, and some Elsevier titles. Number two is Iran, with many journals but few repositories and very few open access documents referenced by BASE.

Sub-Saharan Africa: Few journals and a small number of documents in open access, but a number of repositories that is above the average. There are some large countries with a growing investment and interest in open access, particularly the green road, like Kenya or Nigeria, but the most important country is South Africa which is also the privileged partner of the Brazil-based SciELO initiative for the development of open access journal publishing on the African continent[3].

Europe & Central Asia: Only one country in this region is considered part of the Global South, Turkey, with a significant and growing number of open access documents, journals (mostly published by universities) and repositories. Compared to other countries from the Global South, Turkey is somehow atypical, nearer to medium-sized European countries than to neighbouring countries like Iran, Egypt or India.

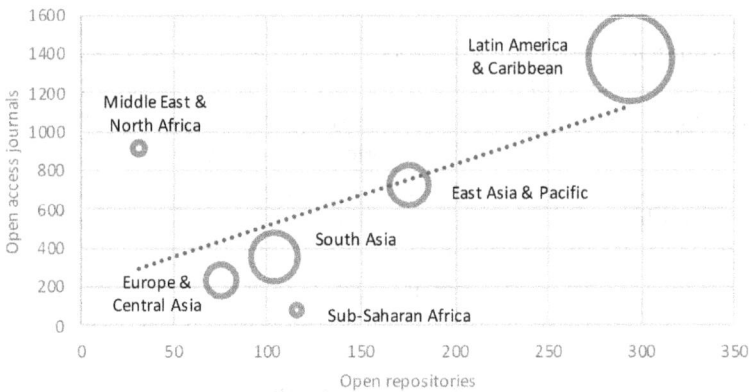

Figure 9. Open access in the areas of the Global South (size of bubble = documents in open access)

These average data are interesting; yet one must keep in mind that more than half of the Global South countries do not contribute in any way to the open access movement—no journal, no repository, and no document

---

3   See Raju, R., J. Raju, & I. Smith (2015). South Africa: The role of open access in promoting local content, increasing its usage and impact and protecting it. In (7), pp. 160–89.

in BASE. Or at least, their contribution is neither visible nor indexed in the main international discovery tools and directories. The comparison with the Scopus database is startling: in fact, only one country (Niger) had no referenced document in 2016, and the median for all Global South countries is 281, which means that all these countries have some kind of academic output, which is visible only on the level of individual authorship and the institutional affiliation.

Income level

Another way to evaluate the development of open access in the Global South is through analysing income levels. The World Bank Group distinguishes four different groups, based on the 2015 gross national income—low income ($1,025 or less), lower middle income ($1,026–4,035), upper middle income ($4,036–12,475) and high income ($12,476 or more) (Figure 10).

	1	2	3	4	5
High income	9	90739	109261	54	119
Upper middle income	31	740916	1649205	468	2028
Lower middle income	34	222225	834884	245	1525
Low income	24	8589	4403	28	13

Figure 10. Open access in income groups of the Global South (1 countries, 2 publications 2016, 3 documents in open access, 4 repositories, 5 journals)

The main contribution to open access is not provided by the high income countries like South Korea or Chile but from the Upper middle income group, which includes large countries with an important research sector such as Brazil, Argentina, Mexico, China, Turkey, Iran or South Africa. Their average share in open access output is two times higher than for the Lower middle income countries, which includes Egypt, India, Indonesia, Sri Lanka or Pakistan. But this average contribution should not mask the fact that the median is quite low and that half of these countries have no or only one or two repositories or journals.

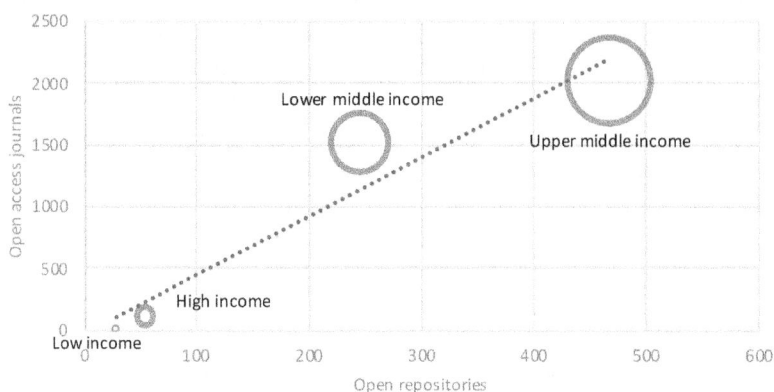

Figure 11. Open access in the income groups of the Global South
(size of bubble = documents in open access)

Rounding out the picture, low income countries (mainly Sub-Saharan countries but also Nepal and Haiti) with a small or nonexistent academic production are more or less absent from the open access landscape. Some exceptions like Tanzania, Senegal, Zimbabwe or Mali confirm the rule. Figure 11 illustrates the differences between these World Bank categories.

Profiles

Finally, the data of OpenDOAR, DOAJ and BASE allow the description of some particular profiles of those countries with some kind of contribution to the open access movement (N=41). We can distinguish two larger groups (Figure 12):

Q1—A leading group with 16 countries which account for 79% repositories, 71% open access journals and 93% freely available documents (BASE). These countries have in common the fact that their repositories and journals are above the median of the Global South. In this category we find all large and significant open access countries such as Argentina, Brazil, China, India, Indonesia, Mexico, South Africa and Turkey, and also countries with less impact and visibility like Bangladesh, Chile, Ecuador, Malaysia, Peru, South Korea, Thailand and Venezuela.

Q3—A second group with 15 countries where the open access movement exists but seems less developed or just at the beginning; together they represent less than 5% of the open access in the Global South. Here there are many countries from Latin America and the Caribbean, like Nicaragua or El Salvador.

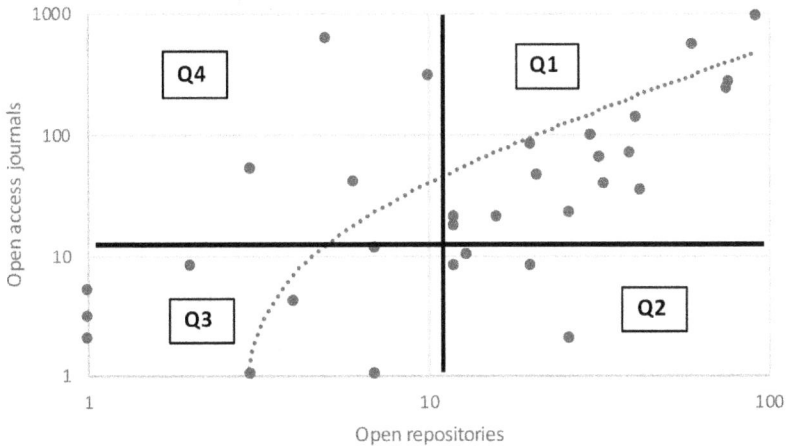

Figure 12. Open access journals and repositories in the Global South (only countries with open access contribution, N=41; with median; logarithmic scale)

Two other groups are less representative and composed of rather atypical countries.

- Q2—Four countries with more repositories and less journals. Together they represent 10% of all open repositories but less than 1% of all open access contributions. Algeria is part of this group, as well as Sri Lanka.
- Q4—Five countries with more journals and less repositories. Their journals represent 27% of all open access journals in the Global South; yet the share of their freely available documents does not exceed 4%. Egypt belongs to this group along with Iran and on a much lower level, Pakistan, Costa Rica and the Philippines.

Among the World's poorest, heavily indebted and least developed countries[4] only four appear to have developed some significant and visible open access activities. These are Kenya and Nigeria with over 20 open repositories each and Pakistan and Bangladesh with a combined 77 open access journals.

23 countries are considered to be fragile or conflict affected; none of them contributes to open access, except for Zimbabwe with ten repositories indexed by OpenDOAR. No small country (Botswana, Djibouti, Fiji, Jamaica, Tonga etc.) is visible or makes any impact.

Lessons learned

Which lessons can be learned from this overview? Let's summarize the current scene in three points.

Diversity: "Global South" is an umbrella term comprising many different situations, ranging from potential superpowers like Brazil, India or China to small or heavily indebted countries like Djibouti or Bangladesh. What most of these countries have in common, though, is the need for scientific information and the will to foster the impact and visibility of their domestic research results through open access. However, the way to open access can be very different, and we can identify at least four paradigms: successful public initiatives (SciELO in Brazil and other Latin American countries), favourable national policies (India), institutional projects (South Africa) and private entrepreneurship (Egypt). Even if Egypt and Brazil are both significant architects of the gold road to open access, the reason is completely different if not opposed—visionary private entrepreneurship in Egypt with one for-profit publishing house; and in Brazil a mainly public-funded non-profit initiative for the local (regional) scientific community. Our figures, in particular those from Scopus and BASE, draw attention to another distinction that should be made, that is between open access publishing by individual authors affiliated to institutions in the Global South and open access initiatives and projects on the international and/or national level. This distinction explains for instance why Scopus and BASE contain references

---

4   International Development Association (IDA), part of the World Bank; N=55 in our sample

to open access papers from nearly all countries worldwide, even from poor and heavily indebted countries, while those countries remain invisible and without any impact in open access directories.

Transition: Well developed and stable open access situations like Brazil or Egypt (insofar as "stability" makes sense in the context of open access….) are rare and unusual, whereas "transition" appears to better characterize most of the countries of the Global South—that is, transition from low impact, closed, domestic scientific information to open science. China is one significant example, with a rapid but so far (at least from "outside") rather invisible development of an open access journal publishing market (4). Iran is another example of a country where nearly all journals are funded and published by universities while repositories are just at the beginning, without for instance any efficient dissemination in open access of PhD dissertations. But this may change in the near future because the Ministry of Science launched a national policy in favour of the preservation and dissemination of electronic theses and dissertations, which implies support for national or institutional open repositories. While our paper is not about barriers to open access it is evident that factors like infrastructures, costly publishing models; intellectual property, attitudes and awareness and language and literacy issues (see (3) for African countries) accelerate or slow down the development of open access. Also, the quality of open access resources remains a complex and controversial issue (6).

Resources: The successful development of an open access policy needs IT infrastructures, human resources and financial investment. From a political viewpoint, the open access movement may be considered as a kind of low-budget solution for scientific communication in poor regions and less developed countries; this is true in particular the green road, that is, the development of open repositories. Yet, empirical evidence shows quite a different picture: all significant progress of open access, be they on the green or gold road or both, take place in newly industrialized countries (NICs), with strong political leadership, rapid growth of urban centers and population, a switch to industrial economies, foreign investment, and open markets. Open access may not contribute to a new digital divide, but surely exemplifies it.

In other words (7), the reasons for success may be different from one country to another; but we can identify at least three key factors of success: strong public policy in favour of open access (including copyright law transparency), growing awareness of open access among the research communities (including incentives and the redesign of the academic reward system), and investment—that is, private investment (with an emergent open access market) and above all, public funding for the dissemination of research results.

References

(1)   Audet, R. (2009). *Du tiers-monde au Sud global : le renouveau de l'activisme diplomatique des pays en développement à l'OMC : une analyse du discours et des formes organisationnelles.* Ph.D. Thesis, Université du Québec à Montréal (UQAM).

(2)   Beall, J. (2012). Predatory publishers are corrupting open access. *Nature* 489 (7415), 179.

(3)   Fox, M., and S. M. Hanlon (2015). Barriers to open access uptake for researchers in Africa. *Online Information Review* 39 (5), 698–716.

(4)   Hu, D., B. Huang, and W. Zhou (2012). Open access journals in China: The current situation and development strategies. *Serials Review* 38 (2), 86–92.

(5)   Kindling, M., et al. (2017). The landscape of research data repositories in 2015: A re3data analysis. *D-Lib Magazine* 23 (3/4).

(6)   Nobes, A. (2017). Critical thinking in a post-Beall vacuum. *Research Information* (89), 10–12.

(7)   Schöpfel, J., ed. (2015). *Learning from the BRICS. Open Access to Scientific Information in Emerging Countries.* Sacramento CA: Litwin Books LLC.

Acknowledgments to Behrooz Rasuli from IRANDOC (Tehran) for offering helpful advice.

Postcolonial Open Access

Florence Piron

Is open access the solution to many of the problems faced in the Global South by postcolonial universities, lacking the resources and capacity to subscribe to expensive scientific journals? In this chapter, drawing on an action-research project in Haiti and in Francophone Africa, I argue that this is not the case. On the contrary, open access can become a tool of neocolonialism if it only gives students and academics better access to science from the North. I conclude with recommendations to make open access an instrument of emancipation and cognitive justice in Africa and Haiti.

Introduction

To promote open access, researchers often argue that it leads to an improved and quicker circulation of research results, thus avoiding unnecessary repetition of scientific work as well as generating new ideas, hypotheses and research projects (8)(35). This argument of increased scientific productivity through open access is perfectly compatible with the "imperative of innovation" at the heart of cognitive capitalism. A more general argument in favor of open access is that it contributes to creating a "knowledge society" by giving different audiences direct access to scientific publications. Non-academics and graduates can continue to train and learn, for instance members of public administrations, teachers, professionals, journalists and civil society organizations' members.

The latter argument is directly relevant for the Global South, in particular Francophone sub-Saharan Africa, where universities lack the resources and capacity to subscribe to the expensive scientific journals that are the focus of the attention of researchers in the North. Is open access the solution to many of the problems of science in the Global South? I would first like to show that this is not the case: on the contrary, open access can become a tool of neocolonialism. In the second part of the chapter, I reflect on how to make open access an instrument of emancipation and cognitive justice.

This analysis is based on numerous readings and reflections as well as on a recent action-research project on the barriers to open science and cognitive justice in the universities of Haiti and Francophone sub-Saharan Africa (26). Without my innumerable conversations, virtual and in person, with young scientists from these countries with whom I have published several papers and a book (24)(25)(26)(27), my thinking would not be what it now is. I thank them very much.

Open access: a catch-up tool or a neocolonialism tool?

Open access to scientific publications is an opportunity to accelerate and reinforce the circulation of scientific knowledge in the scientific world and beyond. Could it then allow African and Haitian universities to catch up on scientific documentation and strengthen their research mission, which is at present almost non-existent?

Indeed, as shown by Alperin's map of scientific publications by country (2), Francophone Africa produces less than 1% of the scientific articles in the Web of Science database. Having seemingly failed to mobilize the means to develop the scientific knowledge it needs to support and guide its sustainable development, Francophone Africa continues to depend on research done in the North or funded by the North. Mve-Ondo describes the shock suffered by African universities during structural adjustments and the reduction of public budgets in the 1980s: reduction of teaching positions, lack of science policies, obsolescence of research infrastructures and educational programs, "bogging down" of lecturers who must seek additional income, continual brain drain to the countries of the North, massive increase in student numbers and lack of resources in university libraries (22). Our survey of the academic experience in Francophone Africa and Haiti (forthcoming) confirms this sad situation by showing many cognitive injustices that prevent the development of scientific research activities within these countries. We define a cognitive injustice as anything that can prevent researchers from deploying the full potential of their research capacities in the service of sustainable local development (25). Open access can then easily be seen as a means of catching up, filling the gaps in libraries and helping professors and lecturers stay up to date.

However, this reasoning can only evoke another well-known one, concerning not the scientific development of Africa, but its development in general: "Africa is lagging behind the modern world, which explains its underdevelopment", to brutally summarize this hegemonic conception of North-South relations. Out of charity, the countries of the North feel obliged to "help Africa develop", giving rise to the industry of international development.

This vision has been strongly criticized by thinkers especially from the Global South (6)(7) who consider that this model of development is derived from European modernity and is not universal despite its pretentions (1)(19). It was violently imposed by the West on the rest of the world through colonization, which used the argument of lagging to justify the economic and cognitive exploitation of the colonized continents without which modernity could not have prospered (4). Postcolonial criticism of development considers that this exploitation continued even after independence, hampering the real rise in autonomy of Africa, which remains perpetually assisted and dependent on the North (21). According to this critique, the current economic and social divide between the North and the Global South reflects the impossibility of many countries or communities in the Global South to develop in their own way, that is to say, according to their own norms and values, anchored in their territory and their history.

This postcolonial critique can be applied to the idea of "scientific catching-up" for Africa. Is there only one model of scientific development, that of Western science inherited from colonial modernity, or can one imagine a different African science, oriented towards the concerns of the continent?

Mvé-Ondo recalls the colonial origin of African science (22), its continual subjection to Northern research projects and theoretical frameworks (5), and its tendency to imitate Western science without proper contextualization efforts. This is particularly the case with regards to how universities are structured and function (11), maintaining to this day the use of a colonial language in university education. If we remain in the positivist perspective, according to which "science" is universal, then effectively

African science, defined as science done in Africa, lags behind and must be helped to develop so that it increasingly resembles the science of the North. However, if one adopts the critical perspective, then African science should be African knowledge anchored in African contexts and using African epistemologies to answer African questions, while also using other knowledge from the rest of the world, including Western science if relevant. For this African science to develop, Mvé-Ondo suggests to "move from a westernization of science to a truly shared science" (22 p.49) and calls for an "epistemological mutation", a "modernizing renaissance" of African science at the crossroads of local knowledge and science from the North—perhaps in echo to Fanon's appeal to a "new thinking" for Third World countries (10)(29)(30). However, as long as this mutation does not happen and African science mostly tries to emulate Northern science, it will suffer from an "epistemic alienation" (25) which hampers its flourishing.

This critical perspective also leads us to "locate" the science of the North in a specific historical and geographical context. From this perspective, science is far from universal; it is globalized. Inspired by the theory of Wallerstein (34), such as Keim (13), Polanco (28) and others, I consider that science has become a world system whose main merchandise is scientific publication circulating among many institutions of high economic value including universities, research centers, governments' scientific policies, journals and an oligopoly of for-profit scientific publishers (16).

What constitutes this world system? At its center are the countries where the vast majority of scientific publications are produced, especially the United States, Great Britain and Australia. The semi-periphery consists of all the other countries, whether from the North or from the "advanced" South (China, Brazil, South Africa), which gravitate around this center, seeking to penetrate or imitate it by increasingly adopting the English language as the language of publication and the article as the unit of scientific knowledge.

The periphery ultimately refers to all the countries that are excluded from this system; those that produce little or no scientific publications identified within the Web of Science or Scopus databases or whose research is invisible, notably in Francophone Africa. Recall that Alperin's map (2), far from being a photograph of the general state of science in the world, is an image of the world system that the Web of Science is trying to build and govern through its standards and the 33,000 journals it indexes.

In this context, open access appears as a neocolonial tool because it facilitates and accelerates the access of scientists from the Global South to the science of the North. It thus contributes to intensifying the epistemic alienation of these scientists and the extraversion of science from the South to the North. Indeed, by making the work produced at the center of the world system more accessible, open access maximizes its impact on the periphery and reinforces its use as a theoretical reference or as a normative model, to the detriment of local epistemologies: "The consequences are lecturers in the Global South who quote and read only authors from the North and impose them on their students and university libraries who strive to subscribe to Western scholarly journals that do not deal with our problems " (18). Let us examine two examples, Research4life and Article Processing Charges.

Research4life

Research4life, whose slogan is "Access to Research in the Developing World", is the collective name of four programs in place since 2002: Health Access to Research (HINARI), Research in Agriculture (AGORA), Research in the Environment (OARE) and Research for Development and Innovation (ARDI). Their mission is to "provide developing countries with free or low-cost access to academic and professional peer-reviewed content online" in order "to reduce the knowledge gap between high-income countries and low- and middle-income countries by providing affordable access to critical scientific research." These programs have provided free or low-cost access to more than 77,000 scientific journals in 8,200 institutions in more than 115 developing countries. The beneficiary countries are divided into two categories: countries A (72 countries) and countries B (45 countries). For the category A countries, the program is totally free whereas those in category B must pay a lump sum. There are other "charitable" programs of this type, such as the Low Cost Journals Scheme, also known as Protecting the African Library Scheme within Africa, the JSTOR African Access Initiative and The Developing Nations Access Initiative.

Research4life seems to me a perfect example of scientific neocolonialism, hidden under the guise of a charitable and generous gesture inspired by the ideal of open access. Indeed, there is a great resemblance between Research4life's mission and the colonial and postcolonial conception of development as "catching up", as if the only way to combat cognitive injustice and the scientific divide were to distribute science from the North to the South in a charitable way, free or low-cost. Several aspects of the program's operation show this very clearly.

First, the Research4life research consortium is far from being non-profit or disinterested since it includes the International Association of Scientific, Technical & Medical Publishers and more than 185 scientific publishers looking for markets. Second, the universal libraries of the Global South which participate in the program cannot choose the journals they receive, as the "bouquets of journals" on offer are predefined by publishers in the North. Having no control over the journals they offer their readers, those libraries cannot select them according to their relevance to local issues. Third, this program encourages the continuing dependence of these libraries on an external program designed in the North and highlighting products from the North, which may disappear as soon as its philanthropic vein gets exhausted. Not only does this state of dependence hinder the deep understanding of genuine open access by Global South university librarians (such as the Base-search.net search engine and its harvesting of open institutional archives), but it also leads to absurd situations as I found out during a visit in June 2015 to an African university library participating in Research4life. In a library room, the program had installed two computers providing access to its journals. However, to prevent unauthorized access, these computers were protected by passwords that changed every month! At the time of my visit, the librarians had lost track of these changes and, as a result, Research4life's computers were unusable.

This program, therefore, produces a situation opposite to what it seeks, namely "Access to Scientific Literature is Improving the Livelihoods of Communities Around the World", a phrase that is entirely meaningless in reality. This program primarily improves the opportunities of publishers in the North without contributing to the sustainable empowerment of university libraries in the Global South.

Article-processing charges (APCs) and the Global South

In the North, for many scientists, especially in STEM (Science, technology, engineering, medicine) (3), open access now means, rightly or wrongly, "publication costs requested of authors" (14)(32). Indeed, since the early 2000s (17), several scientific journals have begun to charge authors who provide their articles for free, indicating that this is the price to pay to join open access. If readers no longer have to pay, then the authors, the scientists, could pay, especially given that their career or their desire for prestige require they publish more and more: a new captive clientele is born! Even if, according to two recent accounts, only 28% (14) to 36% (20) journals charge these fees, this bold business innovation seems to bear fruit, since these costs now appear as "inevitable, normal" in the eyes of scientists. For example, in a 2014 survey at Laval University in Canada, half of the consulted scientists seemed to think that all open access journals automatically charged authors.

For African or Haitian academics who have to work in very difficult conditions, none of this makes any sense. Admittedly, most scientific publishers practicing APC charitably offer exemptions to authors from the poorest countries who wish to publish with them. However, given the numerous obstacles to such publications, there are so few such authors that this charity seems hypocritical. It can also disappear at any moment. This kind of open access emerges as a new commercial practice within a flourishing cognitive capitalism. It is based on a lucrative business model that focuses on captive authors functioning in a "publish or perish" mode, and is indifferent to material conditions of intellectual work in African or Haitian universities, which currently occupy the invisible periphery.

Open access as utopia

Throughout our action-research project in Haiti and Africa, those discovering the potential benefits of open access made the following complaint on a regular basis about free access to scientific documentation not otherwise accessible: "how to benefit from it while our access to the web, computer and even electricity is not guaranteed?". The difficult conditions of internet connections in most African and Haitian universities make open access a distant utopia. What is the point of making millions of open access articles visible on the web if they are not also materially accessible? In addition, few people in these universities have the digital skills to find these open-access articles on the web (for many African students this comes down to Facebook, through its Freebasics program). It should be remembered that it is often upon starting at university that Haitian and Francophone Africa students first come across a computer. They catch-up quickly, but they need to acquire many of the habits northern country students develop while at school before they can envisage that open scientific texts on the web can compensate for the lack of documents in their libraries. In the words of a Haitian student, Anderson Pierre, "a great deal of people do not know the existence of these resources or do not have the digital skills to access and exploit them in order to advance their research project."

Under what conditions can open access become decolonial?

Apart from Hall and Tandon (12), few researchers in decolonial studies have deeply thought about the possibility of "decolonizing" open access to make it an emancipatory tool. These researchers are focused on the political and epistemological dimensions of the colonization of the mind (33) and forget to even consider the conditions of publication and dissemination of their own scientific production and its accessibility in the countries of the Global South (23). Together with my African and Haitian colleagues (27), we have devised several ways to make open access a tool for emancipation and empowerment in Africa and Haiti.

First, we to challenge the hegemony of the world system of science centered in the northern Anglophone countries by suggesting that "another science is possible" (24). This alternative science should respond to the challenge of sustainable local development in the North and in the Global South, by being pluri-lingual (available in national languages in addition to colonial languages), open to the ecology of knowledge and the plurality of epistemologies, with an inclusive and non-normative universalism, and, of course, available online in open access under Creative Commons licenses.

Above all, this "other" science explicitly proposes to repatriate the ecosystem of scientific publication within universities, rejecting the mediation of for-profit publishers. By using open source software such as Open Journal Systems, developing support for publishing journals in university presses or libraries or consolidating peer-review management services between several journals, it will be possible to live a scientific life in which the free sharing of articles is "normal", as was the case at the beginning of the 20th century (15).

Next, we propose to give open access the mandate to increase the visibility of science produced in the Global South, in order to create more cognitive justice and greater fairness between the visible and accessible knowledge from the North and the Global South. In that regard, open access must take into account the knowledge from the Global South which does not appear in the Web of Science (or equivalent), but which is valuable and relevant to many contexts where it should be freely accessible. While traveling to West Africa, I was disappointed to discover that that the geographers of Ouagadougou (Burkina Faso) knew the European science on the Sahel better than the work of the Higher Sahel Institute in Maroua (Cameroon). The latter is not online, much less in open access. Indeed, African science can be found less in scientific articles published in journals from the North than in the dissertations and masters theses carried out in the universities of the Global South: "A significant part of the scientific research output from Africa does not find its way into the world's well-established international scientific journals. One part is published in the small number of local journals with often poor distribution and visibility. And the rest is grey literature, i.e. "unpublished information and knowledge resources such as research reports, theses and dissertations, seminar and conference papers (often) produced in limited numbers, and with limited circulation even within the institutions where they are produced" (Chisenga, quoted by 31). Therefore, open access in Africa should adapt to this reality and focus on good-quality institutional archiving, instead of publication in globalized Northern journals.

In addition, Africa's scientific development aid, if needed, should be directed less towards immediate access to journals from the North and more towards the development of digital tools and skills in African universities. This improved digital literacy would allow lecturers, students and librarians to benefit from existing open access databases and also to create and improve local scientific resources such as open archives, open access journals or publishing houses, or to scan and put online past publications. This requires a number of necessary policy actions: access to electricity, web and computer labs on campuses, financial support to local scientific journals, science 2.0 training (blogs, Twitter, Facebook) for academics and librarians and, of course, local research grants to produce more local knowledge. This is why my colleagues and I have been lobbying in favour of African institutional repositories, showcasing theses and dissertations as well as research reports where most of African research is located (9). There are currently only three institutional repositories in sub-Saharan francophone Africa and none in Haiti, according to OpenDOAR (May 2017).

We constantly advocate that African university libraries, if better funded and their staff better trained in digital open access technologies (such as free software for interoperable scientific archiving such as Dspace, Eprints, Invenio or Omeka), could play a major role in locating, archiving and preserving local scientific documents as well as in managing these archives.

We also remind African and Haitian students that they can have other referents or ideals than the "Harvard model". We invite them to discover the scientific and cognitive heritage of their own countries in order to gain confidence in their ability to create knowledge relevant for their communities. We are extremely proud of the birth in Haiti of REJEBECSS (Network of Young Volunteers of the Classics of Social Sciences) in the wake of our project. REJEBECSS members are young social science students involved in the struggle for cognitive justice. They try and convince Haitian researchers and students to put their theses and papers online in open access and also find old books to be scanned and put online in open access. As a result, they have discovered the richness of Haiti's intellectual history, which have been invisible for far too long. These empowered young people, passionate about knowledge and increasingly better skilled at manipulating digital open access tools, show that open access can become an instrument of cognitive justice. What is lacking, and essential, is a collective universal right to the Web.

References

(1) Alcoff, L. M. (2007). Mignolo's epistemology of coloniality. *CR: The New Centennial Review*, 7(3), 79-101.

(2) Alperin, J. P. (2013). *World scaled by number of documents in Web of Science by Authors Living There*. Accessed at http://jalperin.github.io/d3-cartogram/

(3) Björk, B.-C., and D. Solomon, (2012). Open access versus subscription journals: a comparison of scientific impact. *BMC Medicine*, 10, 73.

(4) Connell, R. (2014). Using southern theory: Decolonizing social thought in theory, research and application. *Planning Theory*, 13(2), 210-223.

(5) Connell, R. (2015). Social Science on a World Scale: Connecting the Pages. *Sociologies in Dialogue*, 1(1).

(6) Escobar, A. (2000). Beyond the Search for a Paradigm? Post-Development and beyond. *Development*, 43(4), 11-14.

(7) Escobar, A. (2007). Post-development as concept and social practice. In (Aram Ziai ed.) *Exploring Post-development. Theory and practice, problems and perspectives* (Routledge). London: Routledge.

(8) Eysenbach, G. (2006). Citation Advantage of Open Access Articles. *PLoS Biology*, 4(5), e157.

(9) Ezema, I. J. (2013). Local contents and the development of open access institutional repositories in Nigeria University libraries: Challenges, strategies and scholarly implications. *Library Hi Tech*, 31(2), 323-340.

(10) Fanon, F. (2002). *Les damnés de la terre—tome 1*. Paris: Découverte/Poche.

(11) Fredua-Kwarteng. (2015). The case for developmental universities. *University World News*, (338).

(12) Hall, B. L., and R. Tandon (2017). Decolonization of knowledge, epistemicide, participatory research and higher education. *Research for All*, 1(1), 6-19. https://doi.org/10.18546/RFA.01.1.02

(13) Keim, W. (2010). Pour un modèle centre-périphérie dans les sciences sociales. *Revue d'anthropologie des connaissances*, 4(3), 570-598.

(14) Kozak, M., and J. Hartley (2013). Publication fees for open access journals: Different disci-plines—different methods. *Journal of the American Society for Information Science and Technology*, 64(12), 2591-2594.

(15) Langlais, P.-C. (2015). *Quand les articles scientifiques ont-ils cessé d'être des communs ?* Accessed at http://scoms.hypotheses.org/409

(16) Larivière, V., S. Haustein, and P. Mongeon, (2015). The Oligopoly of Academic Publishers in the Digital Era. *PLoS ONE*, 10(6), e0127502.

(17) Marincola, F. M. (2003). Introduction of article-processing charges (APCs) for articles accepted for publication in the Journal of Translational Medicine. *Journal of Translational Medicine*, 1, 11.

(18) Mboa Nkoudou, T. H. (2016). *Le Web et la production scientifique africaine : visibilité réelle ou inhibée ?* Accessed at http://www.projetsoha.org/?p=1357

(19) Mignolo, W. (2012*). Local Histories/Global Designs. Coloniality, Subaltern Knowledges, and Border Thinking.* Princeton: Princeton University Press.

(20) Morrison, H. (2017). *OA journals study 2016: 65% free-to-publish.* Accessed at https://sustainingknowledgecommons.org/2017/02/22/oa-journals-study-2016-65-free-to-publish/

(21) Moyo, D. (2009). *Dead Aid: Why Aid Is Not Working and How There Is a Better Way for Africa* (First American Edition edition). Farrar, Straus and Giroux.

(22) Mvé-Ondo, B. (2005*). Afrique : la fracture scientifique / Africa: the Scientific Divide.* Éditions Futuribles. Accessed at https://www.futuribles.com/en/base/bibliographie/notice/afrique-la-fracture-scientifique-africa-the-scient/

(23) Piron, F. (2017). Méditation haïtienne. Répondre à la violence séparatrice de l'épistémologie positiviste par une épistémologie du lien. *Sociologie et sociétés*, to be published.

(24) Piron, F., et al. (2016). Une autre science est possible. Récit d'une utopie concrète, le projet SOHA. *Possibles*, 40(2). Accessed at http://redtac.org/possibles/2017/02/11/une-autre-science-est-possible-recit-dune-utopie-concrete-dans-la-francophonie-le-projet-soha/

(25) Piron, F., Mboa Nkoudou, T. H., Pierre, A., Dibounje Madiba, M. S., Alladatin, J., Michel, R. I., Achaffert, H. R. (2016). Vers des universités africaines et haïtiennes au service du développement local durable : contribution de la science ouverte juste. In (26) p 3–25.

(26) Piron, F., Regulus, S., and Dibounje Madiba, M. S. (Éd.). (2016). *Justice cognitive, libre accès et savoirs locaux. Pour une science ouverte juste, au service du développement local durable* Québec: Éditions science et bien commu. Accessed at https://scienceetbiencommun.pressbooks.pub/justicecognitive1/. The project has been funded by OCSDnet (iHub and IDRC).

(27) Piron, F., Tessy, D. R., Dibounje Madiba, S., Hachani, S., Mboa Nkoudou, T. H., Achaffert, H. R., Batana, J.-B. (2016). Faire du libre accès un outil de justice cognitive et d'empowerment des universitaires des pays des Suds. In *Libre accès aux publications scientifiques entre usage et préservation de la mémoire numérique. Numéro spécial de la Revue maghrébine de documentation et d'information, n. 25*, p 91–106. Tunis: CCSD.

(28) Polanco, X. (1990). *Naissance et développement de la science-monde: production et reproduction des communautés scientifiques en Europe et en Amérique latine.* Paris: Unesco.

(29) Santos, B. de S. (2008). *Another knowledge is possible: beyond northern epistemologies.* London: Verso.

(30) Santos, B. de S. (2014). *Epistemologies of the South: Justice Against Epistemicide.* Boulder CO: Paradigm Publishers.

(31) Schöpfel, J., and M. Soukouya (2013). Providing Access to Electronic Theses and Dissertations: A Case Study from Togo. *D-Lib Magazine*, 19(11/12).

(32) Solomon, D. J., and B.-C. Björk (2012). A study of open access journals using article processing charges. *Journal of the American Society for Information Science and Technology*, 63(8), 1485–95.

(33) Thiong'O, N. W. (2011). *Décoloniser l'esprit.* Paris: La fabrique éditions.

(34) Wallerstein, I. (1996). Restructuration capitaliste et le système-monde. *Agone*, (16), 207–33.

(35) Wilder, R., and M. Levine (2016). Let's speed up science by embracing open access publishing. Accessed at https://www.statnews.com/2016/12/19/open-access-publishing/

Open Access Initiatives and Networking in the Global South

Iryna Kuchma

This short study highlights the impact of open access in the Global South. Featuring collaborative open access initiatives in Algeria, Kenya, Myanmar, Nigeria, Nepal, Palestine, Tanzania, Uganda and Latin American countries, it showcases success and describes the challenges that we still face. It also questions the concept of the journal article—perhaps already becoming obsolete—and discusses the growing number of preprints initiatives to speed up the availability of research results. The value of regional journal and repository networks enhancing open access content in Europe and Latin America is also discussed, as well as the impact human networks make in the Global South.

Health research dissemination

Dr. Bessie Mukami is a general physician at Embu General Provincial Hospital, a relatively large teaching hospital in Embu, a town located approximately 120 kilometres northeast of Nairobi towards Mount Kenya. Embu serves as the provincial headquarters of Eastern Province in Kenya and is also the county headquarters of Embu County. "You have, maybe, one doctor to ten thousand people," says Dr. Mukami. And as she speaks, her fingers click through pages of open access medical journals on a laptop. Subscribing to medical journals is very expensive, and it can be difficult for doctors to consult each other because of the long distances between hospitals. Open access is solving one of the biggest problems Dr. Mukami has: "Instead of calling other doctors for information, the information is open and available, and you search for what you really want," she says.

Dr. Gerald Nderitu, the medical superintendent at the same hospital and an experienced surgeon of 16 years, also relies on open access research output to help his patients. "When I have a patient and I am not familiar with their condition, I will go back to the internet and update my knowledge using PubMed Central," Nderitu says about the free full-text archive of biomedical and life sciences journal literature containing more than 4.2 million articles. The Kenya Library and Information Services Consortium (KLISC) trained Mukami, Nderitu and other health workers in Embu General Provincial Hospital on open access. And this was just one out of almost a hundred open access awareness raising and advocacy events that EIFL organized in Kenya, Tanzania and Uganda in partnership with KLISC, the Consortium of Tanzania University Libraries, the Consortium of Uganda University Libraries, for medical students, researchers, journal editors and publishers to improve health research dissemination and maximize its visibility and impact.

Information consumers vs producers

Are most people information consumers more than producers? This is what we often hear about researchers from the Global South. Prof. Jackson Too of the Kenya Commission for University Education questions this statement: "For a long time, developing countries are said to be more of information consumers than producers. This has been aggravated by the fact that, majority of the high impact journals rarely publish research from local authors. There is a lot of research done locally, however very little is done to disseminate the output. Needless to say, the skyrocketing costs of journals are greatly inhibiting the availability of research results. The problem may not be obvious for the few institutions that can afford subscriptions to digital editions of journals, but to the many potential users who do not have access. This has led to duplication of research since there are no avenues of knowing what has been done by other scholars."

"Just like MPESA—a mobile phone-based money transfer, financing and microfinancing service—has revolutionized the financial sector in the country, embracing open access will create a ripple effect in the way scholars provide access to their craft and this will improve visibility of our university staff. Therefore, failure to embrace open access means missing a grand opportunity to improve dissemination, visibility, and impact of research findings,"—says Alice Kande, a colleague of Prof. Too. The Commission has called upon universities in Kenya to endorse and commit to open access through open access policy formulation and implementation.

Embu University, Jomo Kenyatta University of Agriculture and Technology, Kenyatta University, Kirinyaga University, Pwani University, Strathmore University and University of Nairobi have already adopted their open access mandates. Those that came first supported those that followed and we were happy to facilitate this collaboration. "The research we do is supposed to be for the public good. We are being funded by public institutions, donors, and I think it's good to be able to share," explains Prof. Lucy W. Irungu, the Deputy Vice-Chancellor, Research, Production and Extension at the University of Nairobi. There are over 81,770 research outputs in the University of Nairobi Digital Repository. This repository has reached six million downloads, with the most popular article in a recent interval of six months downloaded 17,300 times.

Covenant University in Nigeria was one of the first universities in Sub-Saharan Africa that adopted an open access mandate and open educational resources policy. These policies have improved the University's and researchers' visibility; increased industry recognition and collaborations; improved pedagogy leading to better students and graduate performances and helped to create a lifelong programme to extend the University's reach to would-be students. Covenant University is now recognized as the best private university in the country, the second best university in Nigeria and a pioneering institution in emerging global ideas.

Challenges

It is exciting to see successes like this, but what about the challenges? Open access is still facing many obstacles such as confusion, complexity, copyright issues, and lack of trust and lack of commitment. Prevailing journal level metrics in research assessment and evaluation and academic promotion, lack of will to change from research administrators and lack of governmental policies and political support in the Global South make researchers reluctant to fully embrace open access.

Article Processing Charges (APCs)—a fee that some open access journal publishers charge to recover their costs—are usually exorbitantly expensive to researchers from the Global South, especially if they have to pay in a foreign currency. And these new challenges are coupled with traditional ones such as poor internet access and connectivity, lack of digital skills and language barriers.

Building trust using new metrics for research assessment and evaluation; shortening embargo periods ensuring open availability of publications as early as possible; and reducing APCs are some of the strategies to overcome the challenges we currently face.

At EIFL—a not-for-profit organization that works with national library consortia in over 40 developing and transition countries to enable access to knowledge for education, learning, research and sustainable community development—we respond to these challenges by supporting development and implementation of open access policies and mandates; enhancing open access journals and repositories; embedding open access, open data and open science into young researcher's workflows and changing the way research is assessed and evaluated. There are over a thousand open access repositories in our network; over four thousand open access journals are currently published and over one hundred universities, research institutes and research funding agencies have adopted open access policies and mandates.

Publishing initiatives

Innovative publishing initiatives are another important strategy for open access. It does feel good when the immigration and border control officer at Kathmandu airport asks you about the Directory of Open Access Journals (DOAJ)—a community-curated online directory that indexes and provides access to high quality, open access, peer-reviewed journals—and Nepal Journals Online (NEPJOL)—a service to provide access to Nepalese published research with 11,477 journal articles available in full text.

And I would like to see more countries following Algeria's approach of mandating open access to Algerian journals and providing a free digital editorial and hosting platform—the Algerian Scientific Journal Platform (ASJP) hosted by CERIST—to journal editors.

In the past 15 years we have seen a lot of successes of scholarly community-led open access publishing initiatives in Latin America. For example, Latin American Council of Social Sciences (CLACSO) has been successfully publishing open access journals with no APCs, promoting open access institutional repositories, and contributing to institutional and national open access policies. CLACSO collaborates with SciELO (Scientific Electronic Library Online)—a successful cooperative decentralized platform for electronic publishing of open access scholarly journals. SCIELO originated in Brazil, and now has national focal points in 14 other countries such as Argentina, Bolivia, Chile, Colombia, Costa Rica, Cuba, Mexico, Paraguay, Peru, Portugal, South Africa, Spain, Uruguay and Venezuela. Another partner is Redalyc, a non-commercial indexing service with hundreds of peer-reviewed open access research journals, published by more than 500 institutions from 22 Ibero-American countries (Redalyc stands for Red de Revistas Científicas de América Latina y el Caribe, España y Portugal).

A similar non-profit initiative is underway in Africa. African Journals OnLine (AJOL) is the world's largest and pre-eminent collection of peer-reviewed, African-published scholarly journals. 215 journals out of 521 hosted at AJOL are open access with 75,938 full text articles for download.

Then again, perhaps a journal article is already becoming obsolete? New media forms emerge and publishing platforms evolve. Researchers use the Jupyter Notebook—an interactive computational environment, in which one can combine code execution, rich text, mathematics, plots and rich media—to write research papers and books. Which academic publisher will be the first to accept a Jupyter Notebook as a journal article?

Repositories

Open access repositories embedded into researcher's workflows are very high on my wish list and we already see this happening with Overleaf—an online LaTeX and Rich Text collaborative writing and publishing tool that makes the whole process of writing, editing and publishing scientific documents much quicker and easier. Overleaf is linked with over 20 publishers and preprint repositories such as arXiv—providing open access to 1,237,145 e-prints in physics, mathematics, computer science, quantitative biology, quantitative finance and statistics; bioRxiv—the preprint open access server for biology; engrXiv—the preprint open engineering archive; and SocArXiv, which is dedicated to opening up social science. The newest baby in the preprints repositories family is Open Access India's initiative. AgriXiv—open access preprints for agriculture and allied sciences—launched in February 2017. And SciELO has also just announced its plan for the development and operation of a preprints server—SciELO Preprint—to contribute to speeding up the availability of research results.

ASAPbio—a scientist-driven initiative to promote the productive use of preprints in the life sciences—defines a preprint as "a complete scientific manuscript that is uploaded by the authors to a public server. The preprint contains complete data and methodologies; it is often the same manuscript being submitted to a journal. After a brief quality-control inspection to ensure that the work is scientific in nature, the author's manuscript is posted within a day or so on the Web without peer review and can be viewed without charge by anyone in the world. Based upon feedback and/or new data, new versions of your preprint can be submitted; however, prior preprint versions are also retained. Preprint servers allow scientists to directly control the dissemination of their work to the world-wide scientific community. In most cases, the same work posted as preprint also is submitted for peer review at a journal. Thus, preprints (rapid, but not validated through peer-review) and journal publication (slow, but providing validation using peer-review) work in parallel as a communication system for scientific research."

Could repositories provide peer review, or open peer review more specifically? Technology is already there. Open Scholar with the support of OpenAIRE, coordinated a consortium of five partners—the institutional repository of the Spanish National Research Council (DIGITAL.CSIC), repository of the Spanish Oceanographic Institute (e-IEO), The Artificial Intelligence Research Institute (IIIA) in Catalonia, The Multidisciplinary Laboratory of Library and Computer Sciences (SECABA) in Granada, and a provider of DSpace professional development and services (ARVO)—to develop the first Open Peer Review Module for open access repositories (1, 2, 3). Are repositories in the Global South interested in integrating the Open Peer Review Module?

The Open Peer Review Module for repositories has already been requested by our partners in Myanmar—both at the University of Mandalay and University of Yangon. "Open access has opened up exciting possibilities for people from all over the world who are keen to further their academic work. It empowers academics, but with power comes responsibility," says Professor Dr Thida Win, Rector of the University of Mandalay. University of Mandalay has taken ownership and control of their research publications, adopted an open access policy and is launching an open access institutional repository. The University of Yangon is also launching its open access repository to support and promote open science, a trend that maximizes investments in research by making research outputs freely available to the world, ensuring access and preservation. Open access has special significance in this country that suffered decades of isolation.

Palestine is another special case for us: a divided country where there are restrictions on movement between its two parts, the West Bank and

Gaza. "In addition to opening our research to the world, Birzeit University's open access institutional repository, titled 'FADA', which means 'Space' in English, will facilitate learning for our local academics, researchers and students, who face problems meeting each other resulting from restrictions on movement," says Diana Sayej Naser, Director of Birzeit University Main Library and EIFL Country coordinator in Palestine.

Networking

Beyond national borders, open access repositories are connected through regional and thematic networks to promote open research content globally and substantially improve the discoverability and reusability of research publications and data. Two of the largest open access repository regional networks—OpenAIRE in Europe and La Referencia in Latin America— bring together professionals from research libraries, open scholarship organizations, national e-Infrastructure and data experts, IT experts and legal researchers to demonstrate the value of common guidelines and interoperable technologies. OpenAIRE and La Referencia collaborate to develop services on top of repository content, such as aggregating content from many different sources; linking research publications, research data and software; tracking research outputs according to research projects and funding agencies; gathering all research output in one place and providing a broader research context; monitoring open access policy implementation and an organization's open access output; providing aggregated usage statistics and research analytics; and highlighting research trends.

Powerful networking technologies are already there. But what about people? It is still uncommon to have open scholarship officers in the universities of the Global South. Usually this role is fulfilled by a librarian or a research manager already overloaded with pressing tasks and job responsibilities. But national, regional and international open access initiatives are also able to provide guidance and support. We all agree that open content creates more value than a closed one, and that reusable content is even much more valuable. Our focus is not just access *per se*, but indeed the right to re-use research output as a means of building new knowledge. This right to stay current in research and developmentis important everywhere in the world, not only in the Global South.

References

(1) Arvo Consultores (2016). Open peer review module. https://github.com/arvoConsultores/Open-Peer-Review-Module/wiki

(2) Perakakis, P. (2015). Developing the first open peer review module for institutional repositories. https://blogs.openaire.eu/?p=489

(3) Perakakis, P., A. Ponsati, I. Bernal, C. Sierra, N. Osman, C. Mosquera-de Arancibia, and E. Lorenzo (2017). OPRM: Challenges to including open peer review in open access repositories. *The Code4Lib Journal* (35), 2017-01-30.

Open Science, Open Access: Opportunities for the Global South, or Just Another Trojan Horse from the North?

Elizabeth Mlambo

This chapter draws on critical voices that argue that the Global North uses open access as a Trojan horse seeking to exploit the resources of the Global South by means of gaining greater access to them. There are deep rooted historical and structural inequalities in which the North has assumed the role of 'provider' of funding and ideas while the South is the 'receiver' in an environment with little scope for action. The Global North has used the advantage of its wealth and position to exploit the Global South. The chapter also presents the arguments of critics who perceive open access as an opportunity that has arisen for the Global South which is bedevilled by so many problems. The chapter documents the 'open' approaches that have been adopted by the Global South as well as the Global North. It identifies the challenges that the Global South has faced in relation to open access, benefits that have arisen for the Global South and the different approaches to open access by the Global North and how the North has also benefited from the efforts of the Global South. Opportunities through open access have also arisen for the Global South. Scientists in the South now have the opportunity to contribute to the global knowledge base through participation, thereby reducing the South to North knowledge gap and professional isolation. The Global South now has the means to distribute local research in a way that is highly visible and without the difficulties that are sometimes met in publishing in journals. Open access reduces the great divide between the haves and the have-nots of a scientific world. The author concludes that with the Global North dominating the scene, there is a danger of drowning out scholarship from the Global South if the playing field remains uneven. The chapter calls for paying more attention to popular local systems of knowledge, in which reality is larger than logic.

Lack of access

The evolution of information and communication technologies is upsetting the long standing socio-economic and cultural landscape of human societies. This evolution pushes the boundaries of human action and is often regarded as a panacea of various ills of humanity, particularly in countries in the Global South leading to associated hopes and fears (17). The Global South and the Global North's relationship have always been unequal. Europe and North America have for decades dominated the rest of the world with their academic products and canons of knowledge production and consumption (13). The Global South is limited in its ability to participate in scholarly communication simply because knowledge creation and dissemination are shaped by the practicalities of money and technology. Developing countries not only have inadequate financial resources but also lack the skills and infrastructure to manage deployment of these technologies (19). There is a digital divide which is combined with the increasing cost of knowledge.

Access to research in the Global South is bedevilled by a number of challenges. There is lack of access to subscription based scientific journals since all scientific breakthroughs relevant to Africa are published in non-open access journals whose prohibitive costs are in the order of US$30 per article (18). It is these costs that inhibit exposure of African research scientists to these discoveries and their ability to use the most up-to-date research knowledge to strengthen their research.

Third world scholars experience exclusion from academic publishing and communications. The knowledge of the third world is marginalized or appropriated by the West while the knowledge of the West is legitimated and reproduced. Consequently academic publishing plays a role in the material and ideological hegemony of the West (1). Africa has produced researchers and educators with "little capacity to work in surrounding communities but who could move to any institution in any industrialized country, and serve any privileged community around the globe with comparative ease" (13).

Lack of visibility

Researchers in the Global South are not visible on the international platform. The problem of information access is one of the barriers. There is lack of technical, educational, political and economic resources to guarantee this visibility. This argument was supported by a cited survey which was undertaken on the information gaps in the Global South. Important information from less developed countries is lost to a wider audience. Poor funding—from public and private sources—remains the most serious limitation to fostering a research culture in the South. "Journals are heavily biased towards high-tech lab-oriented medical research. The community methods, such as qualitative data collection, health education, or the use of village health workers to improve outreach, are not viewed as sufficiently scientific" (7). Most of the time researchers in the Global South perceive the cold hand of prejudice from journal editors whose publication policies repeatedly fail to reflect global burdens of disease.

The lack of visibility in African published journals is attributed to the fact that the journals that do exist are not known about and therefore are not used (16). The majority do not have a viable subscription base and lack long term sustainability. Not many are indexed in international indexing and abstracting databases. African scholars are doomed to consume not books and research output of their own production or choice, but what their affluent and better-placed counterparts in North America and Europe produce and enforce (13). A journal without visibility disappears, results in poor impact, and a null level of citation. With poor impact the authors are not willing to contribute good quality articles (6). In short, open access is tilted in favor of the Global North.

Language poses an additional barrier to developing region scholars. Western publishing, especially in science, primarily uses English; however, not all scholars have sufficient command to publish in this language. English is a second language for many African and Asian researchers, learned in school only and not spoken at home. Its entrenched role in producing and disseminating science privileges researchers from the United States and many current or former British Commonwealth nations (1).

According to Diana Rosenberg (16), academic journal publishing in Africa has been going through difficult times for a number of years. University presses have declined and many once-renowned periodicals and journals have ceased publication or been reduced in size and frequency. Research is suffering because the means to publish research results are lacking and the results on which to develop further research are not disseminated. Knowledge about the journals that continue to be published in Africa, despite all the difficulties faced, is very limited. The contents of such journals are rarely included in international indexes, abstracts, and databases (2).

Lack of relevance

Research is not generally accessible in the South. Works that are more easily found will be more frequently cited (3). And yet for the articles published in peer reviewed journals, Czerniewicz notes that USA academics represent 30%, developing country academics only 20% (which are half from China, India, Brazil, Turkey), and academics from Sub-Saharan Africa just 1%. African scholars face a critical choice between sacrificing relevance for recognition or recognition for relevance to conditions in their local communities (13).

Open access for the Global South leaves a vast inequality in the scientific discourse. Non-attendance at scientific meetings, or not being a member of a prestigious institution, can leave scholars frozen out of the cutting edge of scientific discussions (4). With the leading journals and publishers based in Europe and North America and controlled by academics there, African debates and perspectives find it very difficult to receive fair and adequate representation. Furthermore, when manuscripts by Africans are not simply dismissed for being 'uninformed by current debates and related literature', they may be turned down for challenging conventional scholarly wisdom and also by traditional scholarly assumptions about their content (13). This is why a researcher in the Global South can usually only witness the debate after it has been published in a review.

Lack of funding

Most African university libraries are underfunded, struggle to keep pace with the latest publications of relevance, and are often desperately under stocked and at the mercy of donors dying to dump old and outdated publications as a sort of intellectual 'toxic' waste (13). A case in point is the Muhimbili University College of Sciences whose medical libraries are now frequently devoid of journals because of the prohibitive costs of subscriptions which are payable in foreign currency. Some of the medical databases like Medline do not provide full access to journals although they provide abstracts. One scholar cites the *Lancet* (a medical journal) which costs US$375 per year for the college and US$119 for a personal subscription. This means the college is charged at a higher level than the University College in London (12). This implies that the values and practices shaped by the Northern agenda contribute to the imbalance and the Global South is not able to participate on equal terms with the Global North.

The conventional model of scholarly communication, based on journal publication, has failed to make information accessible and usable especially for the developing world owing to toll-based access. Africa, Sub-Saharan Africa to be specific (and Zimbabwe to be precise), has not been spared this predicament, as the toll fees are beyond the reach of many institutions (11).

An uneven playing field

Contemporary scholarship is dominated by some non-Africans who have strategically positioned themselves as the authoritative voices in a 21st century scramble for influence, as if Africa were a *tabula rasa* with no intellectuals or knowledge production of its own. This form of erasure is not only problematic, but also dangerous. It is clear that those who produce knowledge about something wield considerable power over it (14). The open access movement is a case in point. It is important to note that most of the funding and ideas originate from the North while the South is a receiver in an uneven playing field.

There is a structural imbalance between the South and the North and this means that local scholars experience considerable difficulty when they attempt to intersect with international scholarly networks. This difficulty in access to information affects publication and the visibility of research in the Global South (21).

A hidden agenda?

The world's scientific community is heavily dominated by developed countries, whether one looks at resources, the number of researchers, or scientific "production" (9). Some of the challenges, with specific reference to the Zimbabwean situation, include a small and poor publishing industry, brain drain (the emigration of qualified and skilled manpower from Zimbabwe to other countries), technological problems, legal and copyright issues, communication and marketing issues, as well as limited government and institutional support (12).

Some scholars have argued that the Global North has a hidden agenda in promoting open access in the Global South. A growing number of research funding bodies in the United Kingdom, the European Union, the United States have recently adopted policies that all research funded by them must be made open access. These funder mandates will raise the visibility of their research outcomes… and this is likely to bring benefits to countries in the North (10).

According to Elena Šimukovič, *Voices from the Global South* and others, less privileged regions of the world have been overwhelmed by a reinforced dominance of the Global North. The "pay to say" principle has created even more inequalities among scholars and academic institutions. In the global scholarly communications community, the existing power structure of Western scholarship is still being reproduced. This keeps open access content from the Global South in the margins. Consequently, open access projects from the Global South remain impactful primarily at the regional or local level (3).

The difficulties in research, publication, editorial bias, and information access facing the least-developed parts of the world are profound and often seem almost intractable. This is due to the perversity in the world of research that centralizes information in the Northern hemisphere and probably delays useful local application of new knowledge (7). Today most published research reflects Western interests and Western academic thought and will not benefit the local population.

The Global Open Knowledge Hub—A case in point

The Global Open Knowledge Hub (GOKH) is a project funded from the Global North which aims to improve the supply and accessibility of content that supports evidence informed policy making and practice by development actors. It is a technical, content driven program that aims to develop new technical infrastructures and standards to support the increased access and availability of research knowledge.

The British Library for Development Studies (BLDS) holds Europe's largest collection on economic and social research. BLDS also runs a number of global outreach programs including digitising and profiling southern research in collaboration with institutions across Asia and sub-Saharan Africa. GOKH is a project run at IDS by BLDS through which literature from the global south have been collected and deposited into the British Digital Library and made available on an open access basis.

One of its key aspects is encouraging the contribution of content from the South, recognizing that digital divides and prior power imbalances can be recreated within an open data environment. Below are statistics of downloads from one of the sub communities of the British Digital Library, OpenDocs, the institutional repository of the Institute of Development Studies (IDS) with publications by IDS Fellows, partner research centers and consortia and other publications from research institutes in developing countries curated by the BLDS. Most countries from the South are beneficiaries of this project. However, statistical evidence indicates that the Global North is a big consumer of this literature.

Country	File downloads	Item Views	Sum Total
United States	1,168,345	1,342,697	2,511,077
United Kingdom	348,835	390,488	739,323
Russia	337,191	31,472	368,665
Germany	262,112	312,582	574,694
Ethiopia	235,140	27,958	263,098
France	172,829	220,560	393,389
India	131,530	29,380	160,910
Kenya	62,365	15,067	77,432
South Africa	56,429	14,654	71,083

Figure 1. Usage statistics from OpenDocs (retrieved 6 March, 2017)[1].

64% from the total item views (3,593,178) and 42% from the total file downloads (5,505,121) are from the five countries USA, UK, Russia, Germany and France. The above statistics are a clear indication that the Global North is a major beneficiary of research output from the Global South, reinforcing the argument that open access is likely to bring benefits to the North. There is a continued Northern economic advantage at play, which contributes to such imbalances (8).

Opportunities from open access, some scholars argue, have also arisen for the Global South. Scientists in the South now have the opportunity to contribute to the global knowledge base, thereby reducing the

---

1   https://opendocs.ids.ac.uk

South to North knowledge gap and professional isolation. The Global South now has the means to distribute local research in a way that is highly visible and without the difficulties that are sometimes met in publishing in journals, such as biased discrimination between submissions generated in the North and South (2).

The visibility, usage and impact of researchers' own findings also increases with open access, as does their ability to find, access and use the findings of others. Academic communities in poorer countries can take advantage of servers anywhere in the world offering open access services, without the need to set up their own independent servers or maintain them (20). The poorest countries will benefit most from open access publishing, with free access to research information from developed countries (5).

The open access initiative has generated key benefits, most importantly the removal of subscription barriers to published research. This is especially true for communities that suffer from limited visibility for their research output and/or inadequate access to necessary scholarly journals (15).

No choice?

This chapter has demonstrated how the Global South and the Global North's relationship have always been unequal. Europe and North America have for decades dominated the rest of the world. Open access for the Global South reinforces a vast inequality in scientific discourse. The chapter has also highlighted how open access reduces the great divide between the haves and the have-nots of the scientific world allowing anyone, anywhere on the planet with internet access to read all the scientific reports unfettered by prohibitive subscriptions (4). Through open access African researchers can access more research articles than ever before. There is no question that open access is now firmly part of the global knowledge creation and dissemination landscape. African researches can now generate a much more needed visibility and can reach wider audiences. Unfortunately, this interaction still takes the form of North American and European universities calling the tune for the African pipers. With the Global North dominating the scene, there is a danger of drowning out scholarship from the Global South as long as the playing field remains uneven. The Global South has no choice but to comply with the North as it does not have an alternative due to historically rooted and current inequalities.

References

(1)   Bonaccorso, E., et al. (2014). Bottlenecks in the open-access system: Voices from around the globe. *Journal of Librarianship and Scholarly Communication* 2(2), 1.

(2)   Chan, L., and B. Kirsop (2002). Open archiving opportunities for developing countries: towards equitable distribution of global knowledge. *Ariadne* (30), 20 December 2001.

(3)   Czerniewicz, L. (2013). Inequitable power dynamics of global knowledge production and exchange must be confronted head on. *Impact of Social Sciences Blog*, 29 April 2013.

(4)   Dayton, A.I. (2006). Beyond open access: open discourse, the next great equalizer. *Retrovirology* 3(1), 55.

(5)   Faber Frandsen, T. (2009). Attracted to open access journals: a bibliometric author analysis in the field of biology. *Journal of Documentation* 65(1), 58–82.

(6)   Gonzalez-Bailon, S. (2009). Traps on the web: The impact of economic resources and traditional news media on online traffic flow. *Information, Communication & Society*. 12(8), 1149–1173.

(7)   Horton, R. (2000). North and South: bridging the information gap. *The Lancet* 355(9222), 2231–2236.

(8)   Jentsch, B., and C. Pilley (2003). Research relationships between the South and the North: Cinderella and the ugly sisters? *Social Science & Medicine* 57(10), 1957–1967.

(9)   Karlsson, S. (2002). The North-South knowledge divide: consequences for global environmental governance. *Global Environmental Governance*.

(10) Kell, C, and L. Czerniewicz (2016). Visibility of Scholarly Research and Changing Research Communication Practices: A Case Study from Namibia. *Research 2.0 and the Impact of Digital Technologies on Scholarly Inquiry*, 97.

(11) Kusekwa, L., and A. Mushowani (2004). The open access landscape in Zimbabwe: the case of university libraries in ZULC. *Library Hi Tech* 32(1), 69–82.

(12) Nyambi. E., and S. Maynard (2012). An investigation of institutional repositories in state universities in Zimbabwe. *Information Development* 28(1), 55–67.

(13) Nyamnjoh, F.B. (2012). "Potted plants in greenhouses": A critical reflection on the resilience of colonial education in Africa. *Journal of Asian and African Studies* 47(2), 129–154.

(14) Pailey, R.N. (2016). Where is the "African" in African Studies? *Hove Peace Building International*.

(15) Quaderi, N.A. (2011). Open access in 2011: a publisher's perspective. *Retrovirology* 8(2), O25.

(16)   Rosenberg, D. (2002). African Journals Online: improving awareness and access. *Learned Publishing* 15(1), 51–57.

(17)   Sane, I., and M.B. Traore (2009). Mobile phones in a time of modernity: the quest for increased self-sufficiency among women fishmongers and fish processors in Dakar. In *African Women and ICTs: Creating New Spaces with Technology*, edited by Buskens, I., and A. Webb, 107–18. Ottawa: IDRC.

(18)   Siegel, K., D. Mutonga, D.M. Matheka, J. Nderitu, R.D. Alessandro, and M.I. Otiti (2014). Open access academic publishing and its implications for knowledge equity in Kenya. *Globalization and Health* 10 (26).

(19)   Singh, R.K., S.K. Garg, and S.G. Deshmukh (2008). Strategy development by SMEs for competitiveness: a review. *Benchmarking International Journal* 15(5), 525–547.

(20)   Xuemao, W., and S. Chang (2006). Open access–Philosophy. In *Policy and Practice: A Comparative Study Proceedings of the World Library and Information Congress: 72nd IFLA General Conference and Council, Seoul, Korea.*

(21)   Yousif, N.H., and M. Bonati (2000). North and South: bridging the information gap. *The Lancet* 356(9234), 1034–1035.

(22)   Yudkin, J.S., A.B. Swai, and R. Horton (2000). Access to medical information in developing countries. *The Lancet* 355(9222), 2248.

A Tale of Two Globes: Exploring the North/South Divide in Engagement with Open Educational Resources

Beatriz de los Arcos and Martin Weller

In this chapter we consider what evidence exists of a divide between the Global North and Global South in terms of engagement with open educational resources (OER), understanding engagement as the production and sharing of educational materials online. We discuss whether identifying educators as contributors or consumers of OER can be empirically grounded, and advocate advancing internet access in developing countries to reach a global balance where sharing is key.

Introduction

Mainly considered a socio-economic, political and cultural divide, the disparity between the Global North and Global South is also evident in open education: established trends in open educational resources (OER) research originate largely in the US and Europe (13), and the provision of open content and pedagogy tend to be dominated by English-speaking, developed countries (2, 6). It was Albright (2) who first introduced the notion that the world of OER risked being separated into contributors and consumers, if the North was allowed to lead the production of knowledge without reciprocity from the less developed nations of the South. Although some observers later interpreted this as a neocolonial push (3, 5), Albright strongly argued for global balance, for non-English and non-Western settings to be given a voice in the shaping of the open education movement.

This stance is now more widely advocated than it has ever been, mostly thanks to organizations such as the Commonwealth of Learning, who are leading Regional Consultation meetings in preparation for the 2017 OER World Congress, and research initiatives such as the influential Research on Open Educational Resources for Development (ROER4D) project and their work on the use and impact of OER in developing countries. Yet evidence that knowledge flows from North to South and not vice versa to date rests mainly with Cobo's study of the coverage of online open content between 2007 and 2011, in which he concluded that the number of Spanish and Portuguese OER in non-academic platforms (i.e. YouTube), although increasing at a higher rate, remained considerably lower than English OER (4).

The potential of OER has been applauded as "the move from passive consumption of educational resources to the formal engagement of educators and learners in the creative process of education content development itself" (11). The question we pose is 'What do we know about how teachers around the world engage with OER, that is create educational resources and make them publicly available, in order to substantiate the existence of a North/South divide?'

The OER Research Hub dataset

From 2013 to 2016 we worked under the auspices of the William and Flora Hewlett Foundation on the Open Educational Resources Research Hub (OERRH) project to investigate the impact of OER use on teaching and learning. We set out to prove (or disprove) eleven hypotheses that summarised established beliefs of the benefits of using OER; for example, OER widen participation in education, OER adoption brings financial gains for students/institutions, open education acts as a bridge to formal education, and so forth (1). As part of our research, we conducted a global survey to ask educators and learners about what they thought of, and how they used, open content. This dataset is freely accessible online[1] and is, to the best of our knowledge, one of a kind in terms of its geographical coverage: we collected just under 7,700 responses from participants in over 180

---

1   See https://figshare.com/articles/OERRH_Survey_Data_2013_2015/1528263

countries, nearly a quarter of them native speakers of a language other than English. For the purpose of this chapter, we grouped survey entries into two categories—Global North and Global South, following Wikimedia's regional classification[2], seeking to determine whether educators' behavior could identify them as contributors (Global North) or consumers of OER (Global South).

The demographic characteristics (table 1) show a sample which is not entirely dissimilar: more male teachers in the South completed the survey; they are younger in age and less experienced than their Northern peers, but equally highly qualified and in full time employment. The largest and probably most obvious difference can be appreciated in the language they speak: while in the North a sizeable majority are native speakers of English, this percentage is halved in the South where native languages range from Malayalam and Afrikaans to Spanish, Arabic or Portuguese. In addition, it is worth highlighting the contrast in favour of the Global South in the number of educators who teach in work-based training and one-to-one contexts.

	Global South (n=584)	Global North (n=1955)
Gender	58% male 42% female	48% male 52% female
Age	28% 35-44 years-old	26% 45-54 years-old
Educational Qualification	60% postgraduate 24% undergraduate	64% postgraduate 20% undergraduate
Employment Status	67% full-time 16% part-time	61% full-time 23% part-time
English Native Speakers	39%	76%
Teaching Experience	49% >10 years	57% >10 years
Teaching Context	40.1% school 38.9% HE 27.2% work-based training 26.8% 1-to-1 tutoring	36.4% school 31.6% HE 15.6% work-based training 17.8% 1-to-1 tutoring

Table 1. Demographics of Survey Sample

Internet user statistics in 2016 revealed that penetration rates of 28.7% in Africa and 45.6% in Asia were below the world average of 50.1%, and well behind Europe (73.9%) and North America (89%)[3]. Bearing this in mind,

---

2   See https://meta.wikimedia.org/wiki/List_of_countries_by_regional_classification

3   See http://www.internetworldstats.com/stats.htm

we analysed the OERRH dataset to determine where and how survey respondents accessed the internet to also find the South trailing the North: higher percentages of users in developed nations have broadband at home, use a mobile phone or a tablet to go online, and are able to connect to the internet at work.

Unsurprisingly, this has an impact on their ability to perform effectively in a digital environment: more teachers in the South than in the North declared their lack of experience with for example, spreadsheet software, virtual learning environments (VLE) for teaching, cloud-based storage, or recording, uploading and downloading podcasts. The only significant difference way in which teachers in developing countries lead their peers in the North relates to their familiarity with a torrent client to share files, which we speculate is a way of bypassing poor internet connectivity.

Engagement with OER

The OERRH questionnaire covered engagement with OER based on three main statements: 'I have created educational resources for teaching', 'I have adapted OER to fit my needs in the classroom' and 'I have created resources and published them under an open license.' In addition, three further items explored educators' use of repositories, enquiring whether or not they had uploaded resources, and whether or not they had added comments either regarding the quality of a resource or suggesting ways of using a resource. All these can be easily mapped to Wild's model of OER engagement, which employs the metaphor of a ladder to represent educators' journey from unawareness to advocacy, and from using 'free stuff on the web' to fully embedding OER in their teaching—sharing their own materials online with a Creative Commons licence, reusing others' materials, and re-sharing and commenting on new adapted versions of these (12).

Applying Wild's model we find that, despite a peak in those who say they adapt resources, low levels of engagement are present in North and South alike (figure 1). Teachers in developed countries indicate that they create classroom resources, and share them online with an open license marginally more often than teachers in developing countries; however, the latter outdo the North in telling others how they have used a resource and assessed its quality. These percentages are not hugely disparate for us to talk about a divide, let alone brand the South as passive consumers.

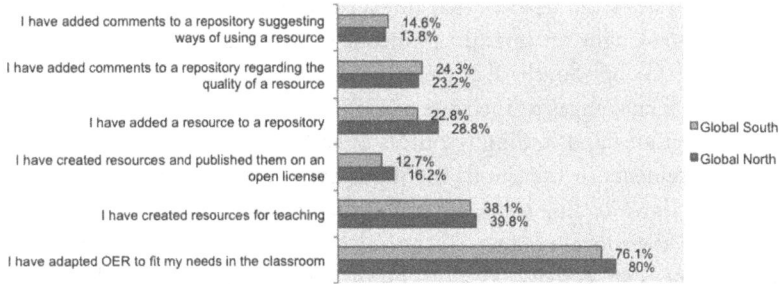

Figure 1. Educators' Engagement with OER

To explain this, we can hone in on what it means to adapt a resource. Okada and colleagues talk about adaptation on an equal pairing with reuse: summarising, repurposing and versioning content, or altering the structure, format, interface or language of a resource (8). Of these, having established that most online content has been created in English, it is obvious that teachers in the South would have to translate material to bring to the classroom more frequently than their northern, mostly English-speaking counterparts.

More importantly, when we examine the reasons why teachers adapt, we discover that while in the North 62.3% of educators say that they use OER because it allows them to better accommodate diverse learner needs in the class, in the South this figure increases to 75.4%. This suggests that teachers regularly tailor content to their students, and in doing so they search for, evaluate and remix resources, which would seem to run counter to the idea of those in the South as passive consumers. Rather this can be interpreted as a willingness to act—perhaps not as ready to share with the rest of the world, but focused on responding to individual learners in their classrooms.

A broken model?

Wild's model of OER engagement has come under criticism for not considering "contextual factors that may either enable or inhibit OER use particularly in development contexts" (10). Even if there is a wealth of resources freely available online, successful engagement with OER will depend on access to digital technologies and having the necessary skills and confidence to employ them (7).

Analysis of the barriers that impact educators' open practices in our sample throws light on several significant differences between the Global North and Global South. Educators in developed countries perceive the most crucial challenges to their use of OER as finding suitable resources in their subject area and finding resources of sufficiently high quality; in contrast, respondents in the South highlight overcoming technological problems when downloading resources, and difficulties finding resources that are relevant to their local context. Technical issues are undoubtedly linked to poor internet connectivity; localization of resources legitimizes adaptation, making a stronger case for teachers in the Global South as engaged and potentially dynamic users of OER. A South African educator explains his take on the obstacles that prevent more ample use of open resources as follows:

"In [my subject] the resources I can find are too Europeanised for the South African context and this is true of many areas. (...) The biggest challenge is a huge lack of IT capability amongst the teachers—I ran a workshop for 30 teachers, not one had ever used [a spreadsheet]. Internet connectivity in very rural areas is both expensive and difficult to access, often breaking transmission so you have to restart downloads. Very rural areas need resources they can download and store on a hard/flash drive. Breadbin / granary type models need to be set up in key villages with open access to all. Whilst I can and do use open sources extensively, the teachers I support are unable to."

How do these factors affect teachers' development of educational content and sharing of resources openly online? Perryman and Seal, in their study of OER users in India, observe that educators who experience a high incidence of inhibitors (i.e. slow internet connection, no internet connection or no access to a computer) also show high levels of engagement with OER (10).

	Internet access at work	
	Yes	No
I have adapted OER to fit my needs in the classroom	76.2%	74.6%
I have created resources for teaching	40.4%	31%
I have created resources and published them on an open license	16%	4.2%
I have added a resource to a repository	27.5%	10%

Table 2. Impact of Inhibitors on OER Engagement in the Global South

Our analysis, however, reveals a different picture; comparing the self-reported behavior of teachers in the Global South, unable to access the internet at the place where they teach, against those for whom access is not an issue, we find that inhibitors have a negative impact (table 2). While their ability to engage in the adaptation of the resources is hardly affected, any intention educators might have of sharing these materials beyond the confines of a walled space and for the public is rendered close to futile without the means to do so.

Impact of using OER

If we prioritised advancing internet access in developing countries, would engagement with OER increase? Analysing survey responses to explore the difference between Global North and Global South in how educators perceive the use of OER influences their teaching, we note that overall impact is felt more strongly in developing nations, the largest gap referring to having improved ICT skills, using more culturally diverse resources, having more up-to-date knowledge of their subject area, and covering the curriculum more broadly (figure 2). Do these impacts only happen when teachers have access to the internet? Certainly not, but it raises the question of how much this could be improved with enhanced access.

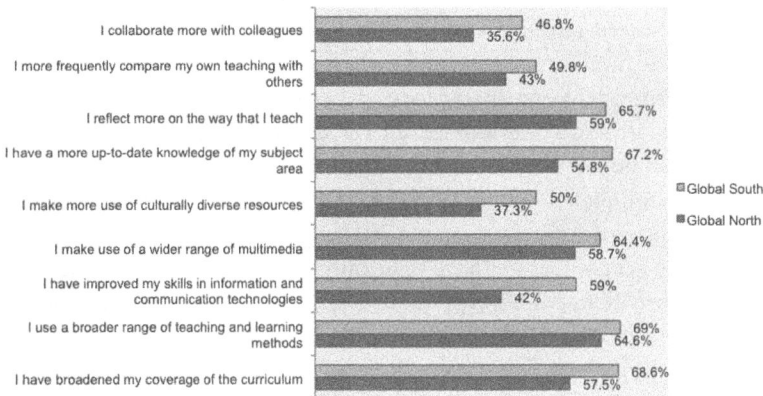

Figure 2. Impact of OER Use on Teaching

Conclusion

Our analysis of the OERRH dataset has provided evidence of the extent to which educators in the Global North and Global South engage with OER. Interpreting engagement as developing content and sharing it publicly online, we found little or no empirical grounding to anchor a North/South divide, and by extension the identification of North as contributors and South as passive consumers of OER. Where a difference exists, this is caused by a considerable difference in access to the internet, which impedes teachers' free use of knowledge. If developed and developing nations had equal access, the argument would likely shift away from access understood as lack of available resources, to lack of suitable resources—i.e. most urgently resources in particular subject areas and of sufficient high quality for a teacher in the North, and relevant to the local context for a teacher in the South. This is a challenge educators in the North and South could address as equals.

Millions of educators around the globe create their classroom materials from scratch or produce versions of content already at hand on a daily basis; if these were regularly shared online with an open licence, there would be plenty of resources out there to choose from in any language, any class level, with examples that are familiar to local students, to teach Mathematics, Philosophy, Economics or any other discipline. This is an area in which we can say quality is in the eye of the beholder; what one teacher cannot use, another will find invaluable; if not good enough as is, an openly shared resource can be added to, undone, built upon, deconstructed and improved so that it works where and how it is needed. It is this sharing behavior that needs to be fostered, facilitated and celebrated by North and South alike, for the discussion to become an honest reflection on how much we are taking and how much we are giving back, instead of an imbalanced and negatively laden assumption regarding roles in open education.

Open access does not have to be only about the dissemination of scientific information, but also about the availability of learning and teaching resources. If the internet can accelerate research, it can also drive teaching and learning. The more educators there are who can access, use and build upon other educator's teaching resources, the more valuable those resources become and the more likely students around the globe will benefit.

References

(1)   de los Arcos, B., R. Farrow, L.A. Perryman, R. Pitt, and M. Weller (2014). *OER Evidence Report 2013–2014*. OER Research Hub, The Open University.

(2)   Albright P. (2009). Discussion Highlights. In S. D'Antoni and C. Savage (Ed.), *Open Educational Resources Conversations in Cyberspace*, pp.61–81. Paris: UNESCO.

(3)   Bateman, P., A. Lane, and R. Moon (2012). Out of Africa: A Typology for Analysing Open Educational Resources Initiatives. *Journal of Interactive Media in Education*, 2(11), Art.11.

(4)   Cobo, C. (2013). Exploration of Open Educational Resources in Non-English Speaking Communities. *The International Review of Research in Open and Distance Learning*, 14(2), 106–28.

(5)   Glennie, J., K. Hardly, and N. Butcher (2012). Introduction: Discourses in the Development of OER Practice and Policy. In J. Glennie, K. Harley, N. Butcher, and T. van Wyk (Ed.) *Open Educational Resources and Change in Higher Education: Reflections from Practice*, pp.1–12. Vancouver: Commonwealth of Learning.

(6)   Hatakka, M. (2009). Build it and They Will Come?—Inhibiting Factors for Reuse of Open Content in Developing Countries. *The Electronic Journal of Information Systems in Developing Countries*, 37(5), 1–16.

(7)   Lane, A. (2009). The Impact of Openness on Bridging Educational Digital Divides. *The International Review of Research in Open and Distance Learning*, 10(5), 1–12.

(8)   Okada, A., A. Mikroyannidis, I. Meister, and S. Little (2012). Colearning–Collaborative Networks for Creating, Sharing and Reusing OER through Social Media. In *Cambridge 2012: Innovation and Impact–Openly Collaborating to Enhance Education, Cambridge, UK, April 16–18, 2012*.

(9)   OECD (2007). *Giving Knowledge for Free: The Emergence of Open Educational Resources*. Paris: OECD-Educational Resources Centre for Educational Research and Innovation.

(10)   Perryman, L.A., and T. Seal (2016). Open Educational Practices and Attitudes to Openness across India: Reporting the Findings of the Open Education Research Hub Pan-India Survey. *Journal of Interactive Media in Education*, 15(1), 1–17.

(11)   Rossini, C. (2010). *Green-Paper: The State and Challenges of OER in Brazil: From Readers to Writers?* The Berkman Center for Internet and Society at Harvard University, 74 p.

(12)   Wild, J. (2012). *OER Engagement Study: Promoting OER reuse among academics. Research report from the SCORE funded project*. The Open University and University of Oxford, 43 p.

(13)   Zancanaro, A., J.L. Todesco, and F. Ramos (2015). A Bibliometric Mapping of Open Educational Resources. *The International Review of Research in Open and Distance Learning*, 16(1), 1–23.

Ubuntu: a Social Justice Pillar for Open Access in Sub Saharan Africa

Reggie Raju

The points of crossover between Ubuntu and social justice have been the unheralded pillars of open access. Both concepts advance the eradication of information poverty and information unfairness. The open access movement is guided by the principle that access to information, an absolute necessity for any level of growth and development, must be made available free to all end users. Social justice challenges structures that perpetuate poverty and injustice through the eradication of information poverty and injustice and open access is the conduit used for this eradication. Ubuntu on the other hand, is a Zulu word advancing communal justice en route to promoting an egalitarian society. The principles of fairness and justice underpin both Ubuntu and social justice. Academic libraries, be it from a Global North perspective (social justice) or from an African perspective (Ubuntu), have been rolling out open access services to ensure information is made freely accessible to the widest reading audience possible. Some academic libraries are now offering a 'library as a publisher' service to take scholarly information to all parts of the global village. Improved access to information will ensure that all sections of the global village can contribute to the growth and development of the whole village.

Introduction

One of the most significant issues influencing the discourse on open access is its benefits. For many years significant attention has been given to the benefits of freely sharing scholarly information with the largest reading audience possible. Although the openness movement has very altruistic underpinnings, it was marketed to researchers as a process to improve downloads and citation. The current method of acknowledging excellence in research is via impact factor and indices; hence, the marketing strategy of using the increase in downloads and to get buy-in from researchers and administrators. Aulisio's (2014) comment, it is "admirable that scholars choose to give up their rights so that their work can be widely read, but it is an excessive approach that ultimately is only benefitting publishing companies and indirectly harming authors and information users" (1) brings perspective to the drive for openness. Researchers should be guided by the need to share for the growth of research and to find solutions to research challenges.

The concepts of Ubuntu and social justice point to the absolute fundamentals of the openness movement. Although most librarians have not openly declared their support for Ubuntu or social justice, they have for many years championed the societal value of openly making available scholarly information. Further, they have been active advocates for the acceleration of scholarly communication programs as one possible solution to the 'university library acquisitions budgets' crisis in which costs of subscriptions have been spiralling out of control. As pointed out by Neugebauer and Murray (2013), "in true librarian ethos, openness has been adopted by the fraternity as a significant movement towards its commitment to societal value" (9). This commitment is confirmed, maybe for different reasons, by Heller and Gaede (2016) who state that librarians must understand their work in the context of social justice, "lest they become complicit in unjust scholarly communication systems" (4). This author holds the view that the librarian's contribution to an egalitarian society using the openness movement is rapidly growing.

The author also posits that there is clear correlation between social inclusion, social justice and Ubuntu. He further argues that one of the mediums for advancing this connection is the openness movement. Before engaging in a discussion on the golden thread that, in the opinion of the author, should be the founding principles of openness, it is prudent to explore the benefits of open access.

Pragmatism versus altruism

Barrier-free access to scholarly information is a necessary component of a higher education system especially for those researchers in developing countries who have limited access to much of the scholarly literature published in subscription journals due to exorbitant subscription costs (1).

Many librarians champion the openness movement as it is their belief that making scholarship and information freely accessible to all users is in line with their ideals and professional obligations as librarians. This professed obligation is confirmed by the International Federation of Library Associations and Institutions (IFLA) which states as a federation representing the sector, that it is committed to the principles of freedom of access to information and the belief that universal and equitable access to information is vital for the social, educational, cultural, democratic, and economic well-being of people, communities, and organizations (1).

It is in the context of the aforementioned obligations that the author engages in discussion on pragmatism versus altruism. For many decades researchers were prepared to 'donate' their intellectual output to large publishing houses in exchange for visibility. Exacerbating this 'generosity gesture' is the growth of one of the 'recognition' processes, namely the impact factor. The pursuit of improved visibility and accessibility, and the ultimate result of an increase in downloads and citation counts, gives legitimacy to the impact factor. As much as this legitimacy is challenged, the author accepts that downloads and citation are the current standard used to affirm the contribution of research output. It is generally argued that acknowledgement by peers via citation is the *de facto* yardstick for measurement of scholarly contribution.

Heller and Gaede (2016) postulate that traditional ways of assessing the return on investment of open access initiatives and institutional repositories is based on pragmatic measures such as download counts and citation (4). Open access advocates have fallen into the trap of using pragmatic measures to promote openness. With all good intentions, these advocates utilized this pragmatic approach but have overlooked the powerful altruistic impact of improving access to critical content to international and/or marginalized communities.

The assessment of open access initiatives, in the view of Heller and Gaede (2016), must exist on a continuum between purely altruistic and purely pragmatic considerations (4). Assessments of open access institutional repositories generally have focused on citation advantage, since these are quantifiable and may have a direct benefit to the institution as it showcases the institution's research agenda and publishing trends. Confirming this pragmatic measure is research that shows that making an article open access tends to improve its chances of citation (15).

On the flipside, universities with a social justice or socially responsive missions need to ensure that their research output is part of public conversations, as a public good. Open access to scholarly output must be seen as contributing a vital academic commodity predicated on its 'social good' status. In reclaiming their role as disseminators of scholarly content for democratic discourse, librarians need to strongly advance the principles of Ubuntu and social justice; social justice and moral obligation must become explicit drivers of open access. Raju, Claassen and Moll (2017) advance the argument that open access should be driven purely for altruistic purposes and those institutions that are relatively advantaged have a moral obligation to share their output and not to use open access platforms simply to grow their profiles—the emphasis should be on development (10).

Censorship in the midst of open access

Many consider the traditional model of publishing to be aligned with censorship. Even if not by design, the exorbitant cost of scholarly information controls the circulation of information and ideas within the global village. Then there is a second level of censorship, namely, economic censorship. The emergence of five big publishers has resulted in almost monopolistic control of the distribution of the one commodity that grows in value with use, that is, information. As these publishers grow and take on more open content via article processing charges (APCs), the noose tightens. The move of these publishers into the open access arena continues to marginalise the Global South. Under the guise of gold open access, the cost of publishing an article via the APC model relegates the Global South to the furthest point in the periphery of the world's knowledge production. These exorbitant APC costs are not only significant for Global South institutions, but they also have a negative impact on Global North institutions with sizeable research budgets.

The altruistic philosophy of openness has been co-opted by big publishers offering, through pragmatic measures of openness, greater visibility and accessibility to the scholarly works of researchers.

Open access within a social justice paradigm

As may be seen from the preceding discussion, the need to migrate from pragmatic to altruistic measures is of significant importance for the openness movement, especially for institutions in the Global South. The author posits that the library's transition into the openness movement was fuelled by the serials crisis and the underpinning motive was located within the social justice paradigm; the inclusion of the marginalized into the knowledge economy was an important driver.

The author proposes that social justice be viewed as the founding pillar of open access. This hypothesis draws from the definition which suggests that social justice be viewed as a system that confronts structures that perpetuate poverty and injustice. Given that the open access movement's primary mission and actual practice is to confront information poverty and injustice, then it can be inferred that social justice is an underpinning pillar of open access. This thesis is supported by Singh (2014) who submits that social justice, at its most fundamental level, is grounded in the belief that all people have the same status (13). Social justice is basically a praxis of inclusion in which society provides all individuals with equal opportunities. Further, it is usually conceptualized within the ambit of the redistribution of resources to improve the situation of the disadvantaged. There is further evidence in support of the thesis that social justice is a lived concept that encompasses acts of fairness, equality and justness towards others (17). They go on to point out that social justice is also relevant to those who are perceived to be privileged. Both the under-privileged and the privileged ought to share in the promise of fundamental human rights and the resultant praxis of justice as fair, equitable and equal. Therefore, irrespective of whether one is privileged or under-privileged, as soon as information poverty presents itself, it has to be eradicated and one of the proposed mediums is open access en route to ensuring that there is social justice—a state in which any form of discrimination is eradicated.

The toll on scholarly output, in the current 'distribution of publication' system, restricts those who cannot afford to pay for access to research output, thus marginalising them from the learning process. Further, limited access to scholarly content negatively impacts on research production as information is necessary for the generation of new information. This toll system fuels the exclusionary tendency of the academy and promotes elitism. The ultimate effect of the toll is the systematic exclusion of the 'poor' from accessing and contributing to the research being done in a particular field. The knock-on effect of this exclusionary practice relegates the Global South to the furthest point in the world's knowledge production chain. As much as the author concedes that information poverty includes the privileged, it is, in the main, the communities from the Global South that are strangled by this information poverty (16).

As indicated by Lara-Alecio (2016), and fully supported by the author, open-access publishing, be it in repositories or directly in open access journals, aids in pushing the envelope for social justice. It provides a more level playing field for all in terms of (a) delivering and obtaining knowledge, (b) understanding up-to-date scientific information, (c) increasing upward mobility and employability, and (d) ensuring greater societal gain (5).

An understanding of Ubuntu

The concept of Ubuntu should be viewed against the backdrop of assertions made by Britz and Blignaut (2). They present information poverty as a community's inability, not only to access essential information but also to benefit from it in order to meet their basic needs for survival and development. Ubuntu, as a concept, is underpinned by a sense of community: 'you are who you are because of your interaction with the community around you, if the community thrives then you will thrive'. One of the founding principles of Ubuntu is the embracement of human diversity, dignity and democracy (8). Brock-Utne adds, when quoting Tim Murithi, that Ubuntu highlights the essential unity of humanity and emphasizes the principles of empathy, sharing and cooperation in efforts to resolve common problems (3). Ubuntu encapsulates moral norms and values such as altruism, kindness, generosity, compassion, benevolence, courtesy, and respect and concern for others'.

Ubuntu is a Zulu word and it carries with it a sense of togetherness, a sense of belonging, a sense of common humanity and friendship. Letseka goes on to point out that Ubuntu is an indigenous African notion of communal justice in post-apartheid South Africa (7). This notion of communal justice promotes an egalitarian society thanks to the assurances in the Constitution of the Republic of South Africa.

Keane (7) brings into the discussion the relationship between Ubuntu and access to information when he posits that Ubuntu refers to a way of living and that resources are essential for creating knowledge and developing potential. Hence, research, within the realm of Ubuntu, should be community-centred and community development and individual development are indivisible. In Africa, sharing and reciprocity are norms throughout the continent. However, it is most critical in the area of academia which is essential to overcome the myriad of challenges that the continent needs to overcome. Brock-Utne, when connecting Ubuntu and the growth of education, quotes the father of African liberation Julius Nyerere who advanced the arguments that the African education system should be based on a co-operative endeavour rather than individual advancement and that there was a need to stress the concepts of equality and responsibility (3). This resonates with the Ubuntu philosophy, in which care for others and co-operation are valued more highly than competition and individual advancement.

As indicated above, social justice confronts the structures that perpetuate poverty and injustice. Social justice advances the redistribution of resources to improve the situation of the disadvantaged. The need to eradicate information poverty is strong. Continuing on the same trajectory, Ubuntu is firmly enveloped within a communal perspective—the community is traditionally the village. In an age of the globe being one village, sharing and co-operation have to become commonplace to resolve common problems, and sharing and reciprocity should become the norm. The sharing of information to resolve the challenges of the 'village' are unconditional. Therefore, it is safe to say that Ubuntu, social justice and open access are all part of the same continuum, that is, a continuum towards an egalitarian society—a society that is not compromised by the lack of access to information to meet its development needs.

Librarians' contribution to Ubuntu and social justice

There are a number of ways in which libraries have contributed and will continue to contribute to ensuring that information is accessible to the disadvantaged. One key way in which such a challenge has been dealt with is via open access. Academic libraries have done exceptionally well in creating and managing open access repositories to ensure that scholarly content is discoverable and accessible. In more recent years, libraries have been delivering a new service—a 'library as publisher' service.

The 'library as a publisher' service is especially relevant to the village as the village is in need of material that is 'local' and addresses 'local' imperatives. The fact that the publishing of content that reflects 'local' imperatives is scarce is due, in the main, to the fact that monographs do not generate large enough profits—in fact, given the small markets, such publishing is not economically viable as affordability, on the side of the users, is a major challenge. Given that publishing outlets are limited as the market is very small (not because people do not want to read these texts; in the current financial climate it is simply unaffordable), for the small

African publisher it is financial suicide and as for the big international publishers they have no interest as the profit margins are miniscule at best.

To overcome the economic viability challenge, librarians have been investigating, testing and offering open monograph publishing services. The desire by African librarians to provide this 'social justice-Ubuntu' driven service is highly commendable given they have developed or are developing a new skills set. Skinner et al. (14) comments that "although publishing is compatible with librarians' traditional strengths, there are additional skill sets that library publishers must master in order to provide robust publishing services to their academic communities" (14). The author is not aware of any library school/science programme in Africa that offers a curriculum and/or training to support this publishing service. Staff currently engaging in the provision of publishing services have grown into these positions; skills to deliver these services have been acquired through self-directed learning including learning by trial and error.

The challenges confronting African academic librarians includes, *inter alia*, the quest for inexpensive/free monographs (especially free textbooks) and the quest for decolonized content. In pursuit of decolonized education, the need for access to indigenous content becomes ever more important. Further, African content is essential to support the changing education paradigm and collection development practices that support a decolonized education. This demand for a 'new' focus in collection development must be viewed against the backdrop of collection development practises that have been seriously neglected over the decades. The author is of the opinion that library publishing services would offer some relief in providing relevant collections in African academic libraries.

Library as publisher

The growth of the open publishing movement in Africa must be set into motion by institutions that are relatively advantaged. It is incumbent on these institutions to take the lead in sharing scholarly output to engender and nurture a culture of research at those African institutions that currently produce a low research output. Some South African academic institutions, via their libraries, have stepped up to the plate to make scholarly content accessible. These libraries offer a suite of diamond open access services. The library as publisher is gaining some traction as a mainstream service provided by higher education libraries in South Africa. Of the 26 public higher education institutions in South Africa, six of them are currently publishing over 30 journals. The University of Cape Town has gone a step further and is publishing open monographs/textbooks using open source software. As a leading research university on the African continent, it has a moral obligation to connect with the principles of Ubuntu and social justice.

One of the many services provided by academic libraries, in terms of the 'library as publisher' service, is the distribution of unique Digital Object Identifiers (DOIs). As a registered member of CrossRef, an international DOI registration agency, the library has the capacity to assign a DOI to each article. The DOI validates some level of authenticity, which reinforces the trustworthiness of the journal article and title. Further, CrossRef ensures that the DOIs are harvestable by leading harvesting institutions (10). It is the opinion of the author that the open 'harvestability' of this content goes some way in addressing the issue of social justice—making scholarly content available to all including the rank and file. This is content that has the potential to improve the quality of life and is now accessible by all who can gain access to the Internet.

Open Monograph Press

Diamond open access publishing is gaining traction, nationally and internationally, albeit very slowly. There are a small number of academic libraries internationally that are publishing monographs via diamond open access. The University of Cape Town Libraries is one of those academic libraries that offer a diamond open monograph publishing service.

The publishing of monographs at the University of Cape Town (UCT) is still very much in its infancy. Although the UCT Libraries have published six titles in the last eighteen months, they still consider themselves to be in pilot mode with regard to publishing monographs in an open access forum; two of the six titles are open textbooks (books written primarily for students and intended to be used in the lecture halls). Currently, academics at the University are placing their unpublished monographs on their websites for use by fellow researchers and practitioners. The possibility of using Open Monograph Press (OMP) to convert the websites into published monographs has now become a distinct reality for the University community.

The author posits that open textbooks published and those that are still in the pipeline to be published would address a myriad of challenges including, *inter alia*, spiralling costs of textbooks and the decolonialization of content as they are available online for free and licensed under a creative commons license. A medical textbook, recently published online by the library, has audio and visual clips embedded to assist with medical procedures; both medical students and practitioners now have access to an essential textbook free of financial constraints. Africa is in dire need of ear, nose and throat specialists and this medical atlas provides much needed support to both practicing doctors and medical students. These diamond open textbooks bring considerable relief to both students and people who are in practice.

Through the publication of these open textbooks, the University of Cape Town is also addressing the transformation agenda of the country and the continent as well as the issue of decolonialization of content. These challenges are being navigated under the broader umbrella of the innate principles of open access and social justice.

Conclusion

The principles of social justice and Ubuntu are two sides of the same coin with the only difference being that one is through a western lens and the other through an African lens. Either way, social justice and Ubuntu must unabashedly advance the eradication of information poverty and information unfairness. Open access has at its epicenter a social justice or Ubuntu underpinning and this altruistic movement must be advanced to ensure that no society is still at the periphery of knowledge production unless they choose to be there—being relegated to that position only because of the lack of access to information must be addressed with absolute swiftness to ensure global equality.

References

(1)   Aulisio, G. (2014). Open access publishing and social justice: Scranton's perspectives. *Jesuit Higher Education: A Journal*, 3(2), article 7.

(2)   Britz, J., and J. Blignaut (2001). Information poverty and social justice. *South African Journal of Library and Information Science*, 67(2), 63–9.

(3)   Brock-Utne, B. (2016).The Ubuntu paradigm in curriculum work, language of instruction and assessment. *International Review of Education*, 62(1), 29–44.

(4)   Heller, M., and F. Gaede (2016). Measuring altruistic impact: a model for understanding the social justice of open access. *Journal of Librarianship and Scholarly Communication*, 4(1), eP2132.

(5)   Keane, M. (2017). Science learning and research in a framework of Ubuntu. In Malcolm, C., E. Motala, S. Motala, G. Moyo, J. Pampallis, and B. Thaver (Eds.). *Democracy, human rights and social justice in education*. Johannesburg, South Africa: Centre for Education Policy Development (CEPD).

(6)   Lara-Alecio, R. (2016). Reflections regarding open-access journals and social justice. *Journal of Childhood and Developmental Disorders*, 2(2), 16.

(7)   Letseka, M. (2014). Ubuntu and justice as fairness. *Mediterranean Journal of Social sciences*, 5(9), 544.

(8)   Nafukho, F. (2006). Ubuntu worldview: a traditional African view of adult learning in the work-place. *Advances in Developing Human Resources*, 8(3), 408–15.

(9)   Neugebauer, T., and A. Murray (2013). The critical role of institutional services in open access advocacy. *International Journal of Digital Curation*, 8(1), 84–106.

(10)   Raju, R., I. Smith, P. Talliard, and H. Gibson (2012). Open access: are we there yet?–The case of Stellenbosch University, South Africa. *South African journal of libraries and information science*, Special launch issue, 1–19.

(11)   Raju, R., J. Claassen, and E. Moll (2017). Researchers adapting to open access journal publish-ing: the case of the University of Cape Town. *South African Journal of Libraries and Information Science*, 82(2), 34–45.

(12)   Raju, R. (in press). Library as publisher: from an African lens. *Journal of electronic publishing*.

(13)   Singh, V. (2014). Open source software use in libraries: implications for social justice? *Qualitative and quantitative methods in libraries (QQML)*, Special Issue Social Justice, Social Inclusion, 49–57.

(14)   Skinner, K., S. Lippincott, J. Speer, and T. Walters (2015). Library-as-Publisher Capacity Building for the Library Publishing Subfield. In M. Bonn and M. Furlough (Eds.). *Getting the word out: academic libraries as scholarly publishers*. Chicago, IL: Association of College and Research Libraries.

(15) SPARC Europe (2016). *The Open Access Citation Advantage: List of studies and results to date* http://
sparceurope.org/oaca_table/, accessed June 30, 2016.

(16) Tafuri, N. (2014). Prices of U.S. and foreign published materials. In *Library and Book Trade Almanac 2014*, 424–59. Association of Library Collections & Technical Services.

(17) Van Deventer, I., P. Van der Westhuizen, and F. Potgieter (2015). Social justice praxis in education: towards sustainable management strategies. *South African Journal of Education*, 35(2), 1–11.

Asymmetry and Inequality as a Challenge for Open Access—An Interview

Leslie Chan

How did you get involved with open access?

My interest in open access was preceded by a broader interest in the nature of knowledge production and circulation. This interest began when I was a PhD student in Physical Anthropology, at the University of Toronto in the late 80s. The term "open access" was not formalized at that time. I was doing research on the evolutionary history of macaque monkeys, which live in various part of Africa and Asia. Through my research, I recognized that a lot of relevant work had been done by researchers in Southern Countries, but that the majority of this work was published in journals to which the University of Toronto did not subscribe. This is when I first began asking questions around how libraries decide which journals to subscribe to, and what their criteria for selection actually entails. As an example, I found one highly relevant research institution in India that had been producing important research for upwards of 90 years. But this research was essentially unheard of by the University of Toronto libraries, because it wouldn't have met their criteria for credible research publications.

Was this a problem of dissemination, of indexing or of discovery tools?

It was a combination of all these aspects. Above all, this problem reflected the way academic libraries selected academic journals. At that time, the library acquisition policy depended mainly on subscription agencies, and those journals did not get in. They were not part of the agencies' catalogs. Libraries go for convenience, and what is not in the agencies' catalogs is "out of scope". Another reason was the emerging practice of the "big deals," of bundling large numbers of journals by major publishers, which became the major business model of academic subscriptions. The problem with the big deals was that they ate up a large part of the library budgets, with less and less money for other journals.

This means that academic libraries often ignored journals from the Global South, and if not, that they didn't allocate the necessary budget to subscribe to them?

Right, this was a problem. And then, there were and still are two other reasons. The first one is that academic libraries make decisions on journal selection partly based on usage statistics. Now, journals from the global South usually have a smaller readership than journals from the USA, Canada or Europe. They are in the margins, part of the long tail, compared to journals from the large academic publishers, and are thus disadvantaged by the library acquisition decision mechanisms. A last issue has to do with negative intellectual perception.

What do you mean by intellectual perception?

Historically, institutions, and in particular publishers, from the global North have largely established the quality standards for journals. Things like peer review, citation formats, writing or rhetoric styles, as well as external markers such as the journal impact factor. Confronted with academic journals from countries of the Global South that they are not familiar with, librarians but also scientists, often assume that if these quality markers are absent or not recognizable, than the journals are of lesser or even questionable quality. This assumption is wrong but it continues today.

How did you find a solution? Did you find a solution?

In the early 90s when the WWW had just become easily accessible, we thought it was ideal to use the web as a means for sharing scientific publications, particularly those that had been traditionally neglected. At the time my colleagues Barbara Kirsop and Vanderlai Canhos had already established an online platform called Bioline for exactly this purpose. Their idea was to run a free platform on the web for peer-reviewed journals in biosciences for publishers who may not otherwise have sufficient resources on their own. We would provide the technology and assistance, the publishers would provide the content, i.e. reviewed papers. The goal was to increase the journals' visibility and discovery of neglected research on the web and to promote knowledge exchange.

In the early days of Bioline, the idea of "open access" was not formalized yet, and the journals that worked with us had the option of providing free access or paid access. By 2004, two years after the Budapest Open Access Initiative, for which I was an original signatory, we decided to host only journals that were willing to provide open access to their content and we renamed the platform Bioline International.

Can you please tell us a little bit more about this platform? Did you succeed? What was the result?

We initially contacted several publishers, primarily through our personal networks, and we asked them to put their journals on our platform. Interestingly, established international commercial publishers didn't take the project very seriously. But for other local publishers, it was an opportunity to improve the visibility of their titles, for example, essential knowledge in tropical medicine, infectious diseases, epidemiology, and biodiversity from local and regional research in developing countries and to experiment with what became known as open access. At one point the platform aggregated about 100 journals from more than twenty countries, including Iran, India, Bangladesh, Nigeria, Uganda, Kenya, Brazil, Chile and Venezuela.

What is the situation today?

The platform—Bioline International[1]—still exists, 25 years after its launch in 1993. Over the years, Bioline International became a pioneer in the provision of open access to peer reviewed bioscience journals published in developing countries. It is now a cooperative project involving two principal parties: the Reference Center on Environmental Information (CRIA) based in Brazil provides server hosting, administration, and server development on a *pro bono* basis, and Bioline International at the University of Toronto Scarborough (UTSC) which oversees content management, project development and research, together with the Centre for Critical Development Studies and the UTSC Library.

---

1   http://www.bioline.org.br/

When we first started, we mused that our job was to make our platform obsolete, as we were confident that as the technologies became more accessible, the publishers would take on their own hosting and distribution. This was in fact the case with many of the journals we used to work with, Medknow[2] being a great example. So in recent years the number of journals we host have been dropping. But this has partly to do with limited resources on our part.

Can you describe the Bioline business model?

First I think the term "sustainability model" would be better than "business model" as we are primarily a mission driven initiative. Since its inception, the platform has been supported mostly by substantial in-kind contributions by CRIA (server hosting and administration) and secondarily the University of Toronto Scarborough (office space and administrative support for the management team). In the past, Bioline also received funding support from organizations including the Open Society Institute, UNESCO and the International Network for the Availability of Scientific Publications (INASP). Since 2008, Bioline has moved towards a community support model that calls for broad based support from libraries as well as communities with the mission of knowledge access for all. For example, OCUL (the Ontario Council of University Libraries), a consortium of 21 universities in Ontario, Canada, has been an early supporting member of our community support model, as was the Max Planck Digital Library. Their support allows us to hire work-study students to perform document formatting for online publishing and metadata enhancement of our partner journals.

What was the impact in developing countries?

In the 90s, a group of Indian colleagues led by Dr. D.K. Sahu understood the Bioline platform as an opportunity to revitalize Indian journals in medical sciences. At that time, many locally published Indian medical journals were not in a good shape. Bioline appeared as a model for Indian learned societies, to improve the quality of their publications and increase their visibility on the web (Sahu & Chan 2004). In early 2000, Sahu launched Medknow, an open access platform dedicated to medical journals from Indian institutions. At the beginning, these journals were also available on the Bioline server.

---

2   http://www.medknow.com/

What was their business model?

Medknow started with a mixed business model—they charged the societies for publishing the medical journals, they added some freemium features, and they had revenue from advertising. After a couple of years, the Medknow strategy became sustainable, allowing investment for their own online manuscript submission and peer review system. After three years, Medknow had aggregated more than 60 journals. By 2010, Medknow was publishing over 100 journals, mostly on behalf of scholarly societies from various developing countries. A real success story.

And then?

In a twisted way, Medknow was a victim of its success. After years of steady growth and development, they became a target of European publishers' takeover strategies because of its wealth of content and the growing markets in emerging countries such as China and India. In 2011, Medknow was acquired by Wolters Kluwer, to become one of the largest open access publishers worldwide, with nearly 400 medical associations and societies from India, China, the Middle East, and other growth markets, and more than 400 journals in several medical specialties. It is a success story but also a sad story.

Why is that?

A sad story for different reasons. Of course, many Indian authors are happy with this development, because they are now published in open access by one of the well-known "international" academic publishers. On the other hand, the natively developed Indian company and platform are now under control of the Wolters Kluwer technology and management. With the takeover, the Medknow journals hosted on the Bioline International had to be removed. D.K. Sahu is no longer involved with Medknow, and my repeated efforts to contact him have failed.

This is a sad story also because this is not an isolated case. I have seen journals developed locally in Mexico, Kenya, Bangladesh, and when they were deemed "commercially viable", they were taken over by multinational publishers and the local publishing capacity was destroyed. Far from reversing this trend, open access has actually enabled this kind of take over as open access became another lucrative revenue stream for commercial publishers through the APC route.

Is this inevitable? Or can this kind of "success-driven takeover" be avoided?

I think there are other options. For instance, take the case of SciELO[3] in Brazil, an open access portal for journals published in Brazil. The initiative receives public funding, and they are supported by a public policy which considers scientific knowledge as a public good, not a private property. This is where the failure occurred elsewhere—letting the market control the production and dissemination of academic knowledge. We should take the private business out of the scientific information system, or at least have some regulatory oversight. You have similar problems with privatizing public health care. SciELO is a public enterprise, which I think is a better sustainability model for long term development than a private company like Medknow.

There is the argument that publishing is best left to the market but I don't buy that. Keep in mind that the bulk of research is publicly funded, and publishing and dissemination of that research only represent a tiny fraction of that research investment. Also keep in mind that library budgets are publicly funded; so why not consider public funding for the supply side of academic publishing? I think public funding of publishing systems is better than paying public money to private companies to gain access to what should be publicly accessible. I have argued before that if all research libraries spend 1% of their acquisition budget for innovative infrastructure in the field of academic publishing and open access, the sustainability problem would be solved. But the problem here is not just financial but political as well: who will take responsibility for such a global project? Who will assume the governance of library based publishing? The library? The academic community? Funding agencies? The question remains open, for the moment.

You criticize instruments from the North such as the journal impact factor because they contribute to the invisibility of research from the global South. Can you give an example?

I've been told by editor colleagues that despite repeated requests to Thomson Reuters' ISI for their journal to be indexed in the Web of Science, they were rejected without transparent reasons. It was not entirely clear how the company made decisions on what journals to include, although it was clear

---

3   http://www.scielo.org

that the majority of indexed journals are published by established presses from the global North, with titles primarily in English. One reason for establishing Bioline International was to counter the fact that many journals were excluded from the ISI mainstream indexing system. The SciELO platform was created with the same motivation. The problem with this system was and still is that it belongs to a private company without any institutional control or public accountability. It is a system that renders large parts of the world's knowledge invisible. The private corporations and the instruments they constructed work hand in hand to maintain this situation, and my impression is that this power imbalance is becoming greater and greater. Open access has not disrupted this asymmetry in power structure.

Elsevier's Scopus database made some efforts to include other sources, especially in life and medical sciences. Is this a potential solution—open up the existing scientometric instruments?

I think this is not a real solution, for three reasons: the private control of inclusion and exclusion remains intact, the expanded selection is based on business-driven decisions and the assessment of the commercial potential of given journals, and there is no real motivation to include content from developing countries. Scopus is no solution to reducing the asymmetry of global knowledge either, as it is controlled by one of the most powerful publishers in the world. Its interest is in profit, not promoting knowledge equity. So far as I can see, the rapidly rising quantity of journals from India and especially China are not increasing the diversity of creativity and epistemology in a global knowledge commons. Recently, a French group came up with the term of "bibliographic diversity" to denote the need for intellectual diversity and representations in global knowledge. I think that this is an important avenue to explore.

What do you think of the SciELO initiative for the creation of specific "regional" scientometrics?

Basically you have two options: either you create your own tools and metrics or you conform to Thomson Reuters[4] or Elsevier. SciELO is a good example of the first option. Yet, the irony of SciELO is that in recent years, many journals on the SciELO platform decided that getting and increasing

---

4   Today: Clarivate Analytics

the journal impact factor is their sign of success. I am not sure that this is a good outcome. It is certainly in contradiction to its original intent.

And altmetrics? Could they reduce the asymmetry?

Again, I think that the opposite is happening. While the original concept is to provide "alternatives" to the dominant metric of journal impact factor, the so called "altmetrics" are really more technical inventions by the publishers to drive traffic to their sites. Keep in mind that the term is also a trade name owned by the company with the same name "Altmetric", and it is being packaged as part of many commercial platforms. There are plenty of critiques of why the prevailing system of impact analysis is deeply flawed, and adding more metrics to the flawed system only deepens the flaws.

But this is another example that those who have access and control of sophisticated technologies profit the most from new technologies. They are, more than others, able to take advantage of these tools and infrastructure. Technically, "altmetrics" are also tied into other "standards" like the DOI or ORCID, and those standards are once again set by organizations of the global North. When the "altmetrics" become another de facto standard, either you play the game or you become invisible. You have very little choice. So it is just replacing the impact factor in terms of dominance, but not challenging the underlying assumptions. Here is also where the technology gap and the lack of funding and resources further disadvantage publishers and scientists from developing countries. More importantly, as they don't have a voice in how "standards" get set and adopted, they are further subjected to frameworks of research quality and excellence imposed on them, rather than setting the standard in their own terms and contexts.

Can open access be a solution? Which kind of open access would you promote?

In the past, I was a big advocate of green open access, of self-archiving in institutional repositories or IRs. For instance, I worked together with the Indian Academy of Sciences when they started to develop institutional repositories in the early 90s, and with the University of Nairobi a bit later on. However, more than ten years later, I see little progress. Institutions are mostly still stuck with journals and are concerned with journal impact factors, despite the proliferation of IRs. In part this is because global North publishers have been very effective in selling their branding and prestige message to countries in the South, particularly countries that want to be seen as being part of the "global" conversation of science and technology. So they have been encouraging researchers and higher education institutions to emulate the publishing practices for gaining reputation and prestige. Unfortunately IRs are left out of this.

So are you more in favor of gold open access, i.e. open access journal publishing?

Today, the distinction between green and gold open access makes less sense than before. In some ways the differences are artificial. The so-called mega-journals, with PLOS ONE[5] setting the example, for instance are more similar to repositories than to traditional academic journals. They are essentially an overlay system, which means article publishing on top of broad based repositories, and "journals" or special collections could be created at will. I don't see why institutions couldn't "publish" with repositories, the question is how to take care of quality control through peer review or other procedures that could be overlaid on repositories. These kind of services are being developed, particularly for "pre-print" servers, which again are repositories by another name.

Could the Liège model[6]—mandatory deposit in institutional repository—be a model for the Global South?

In the past, I would have answered "yes". But I have changed my mind over the years. An institutional policy in favor of open access is important. But the Liège model became an auditing exercise. It is all about counting and compliance with the institutional policy and evaluation of research performance. Researchers become tired with auditing. Another reason why I have changed my mind is that the Liège model is specific to a certain type of institution and cannot be implemented in the same way everywhere else. It is not a model for all academic institutions. Contexts really matter when it comes to policy development and implementation. The one-size-fits-all model doesn't work.

---

5   http://journals.plos.org/plosone/

6   https://orbi.ulg.ac.be/

What do you think of the German OA2020[7] initiative, i.e. transforming the subscription-based journals into open access journals with article processing charges?

I think that gold open access with APCs is one of the interesting systems that has gone bad. Fundamentally, it is a commercially driven tool, an opportunity for the commercial publishers to make even more money with scientific journals. OA2020 is an extreme version of that model and again it will tend to favor those institutions who are already in a strong position to take advantage of this model.

Could this model be a solution for the Global South, either for the access to information, or for the dissemination of their own results, or both?

I am extremely concerned about this initiative. There is no money in developing countries for this model. The institutions and countries adopting the OA2020 initiative express very clearly that it is not their problem that scientists from developing countries can publish or not. It is a very selfish attitude, individualistic and even nationalistic. At the long term, this approach is really bad for global scholarly communication as a whole because some content will dominate. In other words, we will see regionalism emerging, and again, countries from Europe and North America will continue to dominate.

Which is or could be the role of private funding agencies, such as the Wellcome Trust or the Gates Foundation?

One the one hand, it is good to see more private funders entering the open access space (though Wellcome Trust has been there from the start). Recently, we see nine funders (including Gates) forming a collective called the Open Research Funders Group[8]. All of these donors have their own agendas and philosophies regarding open access and their missions. However by making their funded research open (while other, peer-reviewed academic material remains behind closed paywalls), these donors are actually becoming *over*-represented in terms of their contribution towards public knowledge, particularly in the areas of health and development. Early in 2017 we saw that Gates announced their own publishing

---

7   https://oa2020.org/

8   http://www.orfg.org/

platform called Gates Open Research[9] which is modeled after the same platform used by Wellcome. The platform in this case was provided by F1000, another commercial provider. Because these private funders are the voices that everyone can easily access (provided they have internet access), their views and findings are more likely to be taken up and perpetuated by local institutions that cannot afford to pay for alternative knowledge sources. This is potentially a big problem in terms of perpetuating certain biases.

In the past, academic publishers made large efforts to facilitate the access to scientific information in countries of the Global South. Is this not enough?

I assume you are referring to programs such as Research4Life[10], which encompasses four programs—Hinari[11], AGORA[12], OARE[13] and ARDI[14]. These programs provide developing countries with "free or low cost access" to academic and professional peer-reviewed content online. I am not convinced that providing this type of access based on the GDP of a country makes any sense. They only make sense from a business point of view, as these are essentially "market segmentation" strategies, ways to extract as much income from low income countries as possible. The publishers won't lose money if the country is so poor they can't afford to pay, but the moment their national GDP exceeds the threshold, they have to start paying. So to me, these programs are just ways to test out the market, to find where the demands are, and then apply sales pressure accordingly. One of my colleagues likens this to drug selling tactics. Give the drug out for free and once someone is addicted, start charging them. Recently one of my associates attended a meeting organized by a related group called Publishers for Development and she wrote about her impression (Albornoz 2017). Essentially not only are publishers pushing a rather outdated model of development, with access to knowledge from the North as being essential for local development, they are repeating the

---

9   https://gatesopenresearch.org/

10   http://www.research4life.org/

11   http://www.who.int/hinari/en/

12   http://www.fao.org/agora/en/

13   http://www.unep.org/oare/

14   http://www.wipo.int/ardi/en/about.html

message that to be legitimate, researchers and journals from the South need to emulate those from the North. Again, the importance of local research and relevance was largely left out. This is not a model for empowerment.

What do you think about open science? Just a new and ephemeral tendency? Or here to stay?

One of the drawbacks of open access was that it was far too focused on the journal article as the primary research *output* and who has access to that output. To me, an important part of open access should be an exploration of alternative ways for communicating research, aside from a traditional, published journal article.

In this regard, I find open science to be a more useful narrative. Open science aims for the *entire* research process to become more open: including the production of the research question, methodologies, through to data collection, peer review, publication and dissemination. In that way, it is easier to look at *who* is participating in these processes of knowledge production and what kind of power they have in a given context. It allows us to be more cognizant of how power is prevalent in systems of knowledge production, and allows us to think of ways to democratize these processes—to make them more collaborative and equitable.

However, while open science is a relatively new concept, we see, more often than not, that those with power in processes of knowledge production are able to take advantage of the open science discourses and use them to their advantage. We are seeing that the framing of competitiveness in knowledge production and knowledge-as-an-economic-engine is reiterated in open science narratives. For instance, it has become popular to hear people say "*data is the new oil*"[15]. The idea is that data can be used to create knowledge that can be used for economic benefit. Of course, this is generally only true for those with the power to access and manipulate this data. Therefore, the idea of 'extractive research' has not really improved within discussions of open science. If anything, it has become more in line with a neoliberal agenda, in many ways.

Can you describe your work at OCSDNet?

The Open and Collaborative Science in Development Network (OCSDNet) is an attempt to understand how the theories and practices of open science play out in various global South contexts. The network is comprised of 12

---

15   See for instance https://www.wired.com/insights/2014/07/data-new-oil-digital-economy/

research teams that span 26 countries in the global South. As an example, there is a team that is using feminist theory to explore indigenous knowledge and climate change in South Africa[16] while another team is composed of natural scientists looking at water quality and local development in Lebanon[17]. What binds the projects together is the overarching question of whether, and under what conditions, open and collaborative science could contribute to the effective application of research towards development objectives at multiple levels—from individuals to institutions, and from the national to regional and global communities.

A key emerging lesson of OCSDNet is that knowledge is being produced *everywhere* and that there are unique traditions of knowing from around the world. These unique realities call into question our focus on cost and competitiveness and redirect our attention back to how research is intended to improve our well-being. By better understanding the different forms of knowledge, we think that we can actually do better science. So it is a way of expanding how we consider science. It is a chance for a richer and more inclusive science. OCSDNet is trying to remind all of us about that richness. Our work and the Open & Collaborative Science Manifesto[18] we have created is not proposing anything new. It's just a reminder of what we already know as a global community, but that has often been neglected.

Another key lesson is that the different projects in the network[19] are revealing that there are many regional differences, in the way that open science can be practiced and understood around the world, all of which is shaped by history, culture and local politics. All of these different cases give us a unique understanding of how openness is *situated* within different contexts. We believe that 'Openness' needs to be understood in its own context, given these rich diversities.

Essentially, we are encouraging a definition of openness as a *process* rather than as a set of strict conditions that always must be met. It is an adaptive and dynamic process, and one that is always changing.

All questions were asked by Joachim Schöpfel.

---

16 https://ocsdnet.org/projects/natural-justice-empowering-indigenous-peoples-and-knowledge-systems-related-to-climate/

17 https://ocsdnet.org/projects/american-university-of-beirut/

18 https://ocsdnet.org/manifesto/open-science-manifesto/

19 https://ocsdnet.org/ocsdnet-projects/

References

(1)   Albornoz, D., 2017. The rise of big publishers in development and what is at stake. In: OCSDNet blog, Sep 25, 2017. https://ocsdnet.org/the-rise-of-big-publishers-in-development-and-what-is-at-stake/

(2)   Sahu, D. K., Chan, L., 2004. Bioline International and the Journal of Postgraduate Medicine: A collaborative model of Open-Access publishing. In: Esanu, J. M., Uhlir, P. F. (Eds.), Open Access and the Public Domain in Digital Data and Information for Science: Proceedings of an International Symposium, March 12, 2003, UNESCO, Paris. The National Academies Press, Washington DC, pp. 58–61.

Bionotes

Beatriz de los Arcos is a researcher in the Institute of Educational Technology at The Open University, UK and Academic Lead for the Global OER Graduate Network. She has worked on a vast range of open education research projects, including OER Research Hub where she led the project's work on the impact of OER use on teaching and learning in K-12. She is a member of the Open Education Working Group Advisory Board.

Leslie Chan is Associate Professor in Media Studies at the University of Toronto. Since 2001, Leslie has been directing Bioline International, a collaborative platform based in CRIA Brazil for open access distribution of research journals from close to twenty developing countries. With Alma Swan, Leslie co-founded the Open Access Scholarly Information Sourcebook (OASIS) and the Global Open Access Map. A Trustee of the Electronic Publishing Trust for Development, Leslie is on the editorial board of Open Medicine, and the advisory board of the Scholarly Communication in Africa Project based at the University of Cape Town. He is also a member of the Research Dissemination Committee of the Canadian Federation of Humanities and Social Sciences. He is interested in access to knowledge, bridging the digital and knowledge divide, electronic publishing, impact of new media on social practices, open access to scholarly literature, and teaching and learning with new technologies.

Jutta Haider is Associate Professor in Information Studies at the Department of Arts and Cultural Science, Lund University, Sweden. Her research interests concern the material and discursive conditions of information and knowledge and the shaping of trust in contemporary digital culture. She has published on open access, scholarly communication as well as digital information infrastructures (search engines, encyclopaedias, social media) and information practices more broadly.

Ulrich Herb is a sociologist and information scientist, and a project manager and scientific publishing expert at Saarland University and State Library (Germany). His focus areas are electronic publishing, science communication & infrastructure, scientific publishing, scientometrics and science research. He publishes regularly in a variety of professional bodies in the fields of Information Science and Science Research. He is also a freelance consultant in the field of scientific information.

Iryna Kuchma is the Open Access Programme Manager for EIFL—an international not-for-profit organisation that works with library consortia in over 40 developing and transition countries in Africa, Asia, Europe, and Latin America. In 2013, she received the Electronic Publishing Trust for Development Annual Award, in recognition of her "efforts in the furtherance of open access to scholarly publications in the developing and emerging countries". Her research interests include: open access, open education, open innovation, open research data and open science.

Elizabeth Mlambo is Sub Librarian at College of Health Sciences at the University of Zimbabwe. She holds a Master degree in Library and Information Science from State University of New York (Buffalo), a post graduate diploma in Education, Management of Training with IPMZ (Institute of Personnel Management, Zimbabwe) and a diploma in Industrial Relations from University of Zimbabwe. She has developed learning modules on Infopreneurship, Management of Special Publications and Management of Electronic Resources, Innovation Management and Information Access and Use for the Zimbabwe Open University Library Science graduates. Her interests are in Gender and ICT and the empowerment of women. She is currently the Chairperson of Women's Action Group Board of Trustees, and is a Trustee with Rava Zimbabwe Rava.

Samuel Moore is a PhD candidate in the Department of Digital Humanities at King's College London. He is researching open-access publishing and the humanities, focusing on the practical and theoretical foundations for cooperative, scholar-led and commons-based approaches to publishing that are embedded in a range of humanistic research practices. He is also an advocate of scholarly openness in general and was a 2013-4 Panton Fellow in Open Data at the Open Knowledge Foundation. He was editorial manager at PLoS ONE from 2009–2012 and is since 2013 managing editor at Ubiquity Press.

Florence Piron is an anthropologist and ethicist, and a professor in the Department of Information and Communication at Laval University where she teaches critical thinking through courses on ethics and democracy. She is the founding President of the Association for Science and Common Good and its open access publishing house. She has also founded *Accès savoirs*, a science shop in Québec. She is interested in the links between science, society and culture, both as a researcher and activist for a science that is more open, inclusive, socially responsible and focused on the common good, which she interprets as the fight against injustice and environmental degradation. She intervenes orally and in writing in a wide variety of environments, within and outside the academic world.

Richard Poynder is an independent journalist who has contributed to wide range of specialist, national and international publications, including the *Wall Street Journal Europe*, the *Financial Times*, the *Guardian* and the *Telegraph*. He was also at one time the editor of *Information World Review*. He has been reporting on the open access movement for seventeen years and today most of his writing appears on his blog *Open & Shut?* There he publishes regular interviews with open access advocates, essays on open access, as well as on-going commentary on the movement. He also moderates the Global Open Access List (GOAL).

Hélène Prost is an information professional at the Institute of Scientific and Technical Information (CNRS) and associate member of the GERiiCO research laboratory (University of Lille). She is interested in empirical library and information sciences and statistical data analysis. She participates in research projects on evaluation of collections, document delivery, usage analysis, grey literature and open access, and she is the author of several publications.

Reggie Raju is the Deputy Director (Research & Learning) at the University of Cape Town Libraries. He has been in academic libraries for more than 30 years. He holds a PhD in Information Studies. He is the author of several publications in peer-reviewed national and international journals, chapters in books and a book publication. His research focus is on research librarianship with an emphasis on open access. He has participated, by invitation, in several national and international forums engaging the issue of openness in scholarly communications. He has served on the Executive Committee of the South African LIS professional body (LIASA) and is currently a member of the Academic and Research Libraries Standing Committee of IFLA. Reggie Raju is a member of the Editorial Management Team of the *South African journal of libraries and information science* in the capacity of Journal Manager.

Joachim Schöpfel is lecturer of Library and Information Sciences at the University of Lille (France), director of the French Digitization Centre for PhD theses (ANRT) and member of the GERiiCO research laboratory. He was manager of the INIST (CNRS) scientific library from 1999 to 2008. He teaches library marketing, auditing, intellectual property and information science. His research interests are scientific information and communication, especially open access, grey literature and research data. He is also an independent consultant in the field of scientific information.

Elena Šimukovič graduated from the Faculty of Communication at Vilnius University and the Berlin School of Library and Information Science at Humboldt-Universität zu Berlin. Currently she is working on her doctoral thesis in the field of Science and Technology Studies at the University of Vienna. There she works to investigate multi-layered transformation processes in the science system along with development and establishment of open access publishing models. More information on her work can be found at http://orcid.org/0000-0003-1363-243X.

Martin Weller is Professor of Educational Technology at the Open University in the UK. He has been at the heart of many of the OU's open education developments. He chaired the OU's first elearning course in 1999 with 15,000 students. He was part of the original team that gained the funding for the OpenLearn project and was the Virtual Learning Environments Director who made the recommendation to adopt an open source solution in Moodle. He is currently Director of the Open Education Research Hub, also an ICDE Chair in OER and is the Vice-Chair of the UK Association for Learning Technology (ALT).

Soenke Zehle is lecturer in Media Theory at the Academy of Fine Arts Saar, co-initiator and current managing director of the academy's *xm:lab–Experimental Media Lab* and of *K8 Institut fuer strategische Aesthetik gGmbH*, the academy's non-profit company for think tank, transfer and training activities. He is interested in collaborative research projects at the interfaces of art, technology, and design.

Index

www.ingramcontent.com/pod-product-compliance
Lightning Source LLC
Chambersburg PA
CBHW060305220326
41598CB00027B/4237